Henna Art

初體驗

認識印度的手繪藝術

Henna Art First experience —
understand the art of hand painted in India

小美 著

印度傳統手繪藝術——Henna，與現代Body Art的跨世紀對話，
獨特手技的東方視覺美感，為了美麗與祝福存在的各式圖騰，
本書從基本步驟開始，結合詳細圖片教學，
以基礎的點線面為起點，帶你體驗Henna手繪的異國風情～

The dialog between Indian traditional body art – Henna and the modern body art form,
using unique techniques to create oriental beauty.
Various designs that symbolize beauty and blessing will be taught to you step by step from the basics.
With detailed graphics and tutorials, you are able to start from the beginning,
and experience the exotic styles of Henna Body Art~

編輯序

與小美老師的初次接觸，是在一個平凡的星巴克下午。

會後的午茶小敘，對文創產業工作者來說，再平常不過，就在大家喝咖啡聊是非時，鄰桌的身影吸引了眾人的目光。

一個全身上下民族風的女生，帶著幾個人，悠閒卻專注的拿著像畫筆的東西，靜靜畫出美麗又帶有神祕感的圖騰，顏料還散發特殊的香氣，更令人驚訝的是——畫在自己的皮膚上！

人體彩繪藝術行之有年，但能在現場欣賞專業畫作，又是這麼特別的方式，大家再也按捺不住好奇，紛紛上前觀賞詢問，於是開始了我們的 Henna 初體驗。

聽老師熱情分享 Henna 的故事與自己的學習歷程，可以明顯感覺到，歷史悠久、具有深刻文化意義的印度 Henna，讓眼前女生的生命發出強光強熱，也因此樂於將自 Henna 獲得的力量分享出去，希望越來越多人認識、欣賞這門藝術。

幾次接觸後，開始進行這本書的拍照與錄影，在 Mehndi 的特殊香味裡，我們進行了一場愉快的藝術洗禮，看小美老師帶著自信又幸福的微笑，畫出一幅又一幅的作品，以及詳細的步驟教學，還不厭其煩的為我們解說各種基礎技法和祕訣，其中關聯到印度婚俗的部分，更是講得生動活潑，原來 Henna 美麗線條的背後，竟然還包括如此多的民俗風情；小小的圖騰，蘊含了文字無法詳述的心意。

Henna 的迷人之處還有哪些呢？使用哪些手技才能畫出美麗圖騰？快翻開書，和我們一起進入 Henna 的藝術世界吧！

Forward by the Editor

The first time I met X., was on an ordinary afternoon at starbucks.

To those working in the Cultural and Creative Industry, it is quite common to have an afternoon tea gathering after a meeting. Just when we were chatting away over coffee, a figure sitting at the next table caught everyone's eyes.

A girl was dressed ethnically from head to toe, along with a couple of people. She seemed relaxed but focused, taking a marker-like tool and quietly drawing beautiful and mystical patterns. A special aroma arose from her pigment, and the most amazing thing was – she was drawing on her own skin!

Body Art has been around for many years, but to actually witness the creation of a professional work is very astonishing. Everyone was so curious and could not help but go up to her and ask questions. Thus, opened our door to the first experience of Henna.

When the teacher shared with us passionately about the story of Henna and her own learning experiences, it was clear to see that the Indian art of Henna with its deep cultural meanings, is what makes this girl shine with passion in life. Therefore, she also puts a lot of effort in the promotion of Henna, in the hope that more people will get to know and appreciate this form of art.

After a few meetings, we started to conduct photo and video shoots. In the extraordinary scent of Mehndi, we enjoyed a delightful experience of art. We are able to see the smile of confidence and happiness on X.'s face while she draws one design after another, elaborating each and every step in detail, and at the same time explaining to us patiently the use of every basic technique and tips. She was especially lively when mentioning the part related to Indian wedding traditions. Now we see Henna's beautiful lines are rich with so much cultural aspects. These little patterns are able to express more meaning than words can ever do.

What else makes Henna so attractive? What kind of techniques can be used to create beautiful Henna designs? Just start turning pages, and come with us into the artistic world of Henna Body Art!

作者序

　　以前總是以為，畫畫是要拿著畫筆沾染顏料，然後畫在畫紙或畫布上，才叫繪畫，後來發現畫筆沾顏料不是只能畫在紙和布上，還能畫在皮膚上，那時是我第一次接觸到人體彩繪。

　　幾年前，在一趟旅行中，無意間發現了另一項人體彩繪藝術，不需要用畫筆，也不需要沾顏料，而是直接把顏料擠在皮膚上的一種手藝── Mehndi，就這樣沉淪到現在……

　　人生真的有許多意外，這「意外」當然不是會害死人的那種，所謂的「意外」是說不打算放在生涯規劃裡的，或是壓根沒去想過的，卻在某個機緣下冒出來，然後就此再也放不掉的，而且這樣的「意外」還能夠把潛能發揮到極致，這是不是該稱為「意外」？

　　雖說畫 Henna 不過就是自己愛，然後再向外發展到其他人身上，慢慢的野心漸漸大了，希望有更多人能夠認識這項純手工的藝術，於是，擺攤，是！擺攤！不是夜市的攤位，而是現今最熱門的創意市集，能夠參與的，能不放過的都去了，為了自己最愛的，也是唯一會作的，我想我是豁出去了！

　　這麼熱愛 Henna 的自己，對 Henna 的最終期望是──別再把 Henna 跟刺青搞混了。真的！

作者簡介：
　　沉迷於印度傳統文化 - Henna
　　迷戀印度的風格獨特
　　除了畫 Henna，貪玩也是一種技能
　　理性看待世物，感性面對人生

Forward by the Author

In the past, I've always thought that drawings could only be done by brushes and paint, and completed on paper or canvas. Later, I discovered that when you have brushes and paint, you can draw on materials other than paper and canvas; skin, can also become a canvas to draw on. That was my first contact with body painting.

In my journeys a couple of years ago, I chanced upon another form of body painting art, which can be done without brushes and paint. This form of body artwork can be done merely by directly applying dyes onto the human skin. It is called Mehndi. I have been deeply in love with this form of art ever since…

There are many incidences in life. These incidences which I am referring to, are those that were not originally in your life plan, or something you have never thought of. They just pop out right in front of you under special circumstances and take root in your lives. These incidents can even help you achieve your full potential. Aren't these just the best incidents you can have in life?

I started taking a personal interest in Henna and promoted them to others. I grew ambitious, wanting even more people to fully know this raw form of art. I started to promote my work by setting up booths every where, not in night markets, but in the hottest venue nowadays – the creativity markets. I grasped tightly onto each and every opportunity possible, to promote the only thing I know how to do, which is also my most valued possession. This time, I am all in!

I, who love Henna ever so deeply, only have one ultimate aspiration, that is – for people to know the difference between Henna and Tattoo. For real!

Introduction to the Author:

She is addicted to the traditional Henna Art culture,
and obsessed with the uniqueness of the Indian style.
Other than the ability to create Henna designs,
playfulness is also one of her life skills.
She views the world rationally, but views life sensually.

推薦序

第一次和小美與 Henna 接觸時，居然有小時候看新東西時眼睛發亮的雀躍，和小美學畫 Henna 時，在意想不到的專注中，原本為壓力煩躁的我，很自然就畫出渾然天成的自我創作，整個人掏空放鬆，感到無比的輕鬆與愉悅！原來，這也是種舒壓的途徑。用自己喜歡的方式認識 Henna，才會有不同的故事。

— Amada ＜ FAIRTON ＞服裝公司主管

第一次與小美相遇，是在敦南誠品門口，看到一身吉普賽女郎味道的小美神情專注的為客人畫 Henna，卻又隨興的與客人閒聊話家常，當下與友人便決定讓畫 Henna 成為聚會完美的句點。

當指甲花與肌膚接觸的那一刻起，就知道離不開了，圖案一點一點成型，不用怕刺痛，想畫就畫，身體就是最棒的畫布，滿足一段時間就想變化的心。

從客人到朋友，小美，妳的熱情感染了每一個妳畫過的客人，請繼續加油唷！！

— Carolyn ＜巴莎美學 Balsa Lifestyle 負責人＞

它是永久的嗎？！不是耶！好特別的圖騰！於是，我停下腳步，於是，在手上畫了各種圖騰！

藉由不同的概念想法，呈現出不同的藝術變化，真的太特別了！我愛上了 Henna，釋放不同的創意，自由解放創作，很有舒壓效果喔！看著單調的皮膚上呈現不同的美感，你會發現它的美好。

走在各大市集上看見一顆爆爆頭！沒錯，她就是小美！請您不妨也停下腳步，親身體驗！

— Jenny ＜ Henna 課程高階班學員＞

Henna，一種古老又現代的圖騰；小美，一個堅持追求完美的人。兩者相遇激發無限的創意美圖。當你有一位直言不諱又會打槍作業的老師，當你的老師擁有無止盡的靈感，創作讓我腦袋數度呈現當機現象，Henna 滿足我想低調又招搖的心，歡迎加入我們的 Henna 世界。

— Lori ＜ Henna 課程學員代表＞

Preface

The first time I came in contact with X. and Henna, I felt that spark that I used to feel when seeing something that made my eyes light-up when I was little. When I was Learning how to draw Henna from X., I was amazingly focused. It came naturally to me and I drew my own unique Henna pattern. The stress and frustration that I used to have had been totally released. I felt so relaxed and happy! I now realize that this is a way to relieve stress as well. One must understand Henna in their own way, to create different stories.

　— Amada <FAIRTON>

When I first met X., it was at the Elite Bookstore on Dunhua South Road. I saw a gypsy looking girl that was very focused in her work, painting Henna on a customer.

I then saw her chat with her customer, just like they were friends, and right then, my friend and I decided to get a Henna painted on us as the perfect ending of that day.

You know you are going to be addicted to Henna right from the very second it touches your skin. You see, the patterns form before your very eyes. You do not have to beafraid that it will be painful. You can get Henna whenever and whereever you want.

Your skin becomes the most perfect canvas. Henna satisfies your ever changing mood and desires.

From a customer to a friend, X., your enthusiasm infect every customer you have had, keep up the spirit!

　— Carolyn <Owner of Balsa Lifestyle>

Is it permanent? No, it is not! Such a unique pattern! So, I stopped, and got all kinds of Henna designs on my hands!

Different art creations can be created with the incorporation of various concepts and ideas, which is really special! I fell in love with Henna for its creativity and the freedom to Create anything, all of which are great ways to relieve stress! When you see such extraordinary creations on what used to be just ordinary skins, you will understand the beauty of Henna.

If you take a stroll in the creativity market, and you see a girl with an afro hairstyle?

Yes, that's X.! I recommend that you stop at her booth just like I did, and see for yourself the beauty of Henna!

　— Jenna <A student in the Henna advanced class>

Henna, a form of art that is ancient and yet modern; X. is a persistent pursuer of perfection.

The combination of Henna and X. has stricken unlimited possibilities in the creation of beautiful patterns. She is a teacher that is direct and not afraid to reject your work and a teacher that possesses unlimited inspirations.

The intensity of creation at many times overwhelms my brain. It is Henna that satisfies my desire to stay low key and be flashy at the same time. We welcome you to join us in the world of Henna.

　— Lori <Henna class student representative>

推薦序

在我人生開始轉變的時候，與 Henna 結緣在台北安和路的街頭，對我而言，Henna 代表著我每個階段的印記，本書的作者就是有這本事，每次的圖騰，都抓準我當下的心情。

這位身材嬌小的女子，有自己的個性，但又不失對人的體貼與愛，這次她要出書，真的非常期待。自己來體會，就會明白，為什麼這麼多人會愛上 Henna。

— Celia ＜至死不渝的 Henna 粉絲＞

Henna 的圖騰多是由捲曲的線條組成，捲捲的線條勾勒出神祕炫麗的「迷人感」；一頭捲捲的頭髮交纏包裹著無數幻想的「創意感」，無論是個性或圖騰，都只由得你在半開半闔的捲捲圈圈中半窺半究，這樣的「神祕感」就是我認識的小美老師！

— 簡兒寶 ＜自由專案管理人 & 鐵花村攤商代表＞

其實，在認真 Google 它之前，我一直以為 Henna 是一個人的名字，是那個抓我的手到她的大腿上放，然後在手上用奇怪的顏料畫漂亮圖案的那個奇特女人的名字。

認真的去 Google 它之後，我才明白，Henna 不只可以名正言順的摸摸小手，它跟繪畫、攝影、雕刻等創作一樣，都是引人入勝的藝術，而且，Henna 擁有相當有意思的經典文化背景，雖然身為男生，但對於 Henna 我是完全沒有任何抗拒就一頭栽進，如果你有興趣，歡迎一起參與！

— 安仔 ＜台東老宿舍手作雜貨服飾頭子＞

2008 年在印度旅行的途中，第一次體驗 Henna 彩繪，回到臺灣後，無意間在網路上看見小美老師繪製 Henna 的照片，被她游走在傳統與現代之間，卻又獨樹一幟的 Henna 圖騰深深吸引，當下很興奮的馬上報名上課。

記得第一堂上課時，小美說，希望離開這個世界時，能躺在畫滿 Henna 的棺木裡，我聽了先是一驚，之後心裡好感動，完全感受到她對 Henna 的熱情與執著。

能追求自己夢想生活的人，是超幸運的事，小美老師在我心中就是這樣幸福的人。

— Sharkaren ＜ Henna 初級班學員＞

Preface

When my life started to change, I ran across Henna on An-He road in Taipei. To me, Henna represents the symbols standing for my phases in life. The author of this book has the ability to capture my state of emotions with each pattern designed on every encounter.

The petite girl has a lot of character, but she is caring and loving at the same time.

When I knew she was going to be publishing a book, I was very excited. If you were to experience it yourself, you would understand why so many people fell in love with Henna.

— Celia <Henna fan till death>

Henna patterns mostly consist of curly lines and create a mystical and magnificent "charm". A head of curly hair that is full of illusion-filled "creativity". No matter what is her personality or designs, it can only be half-seen through the mystical lines of the curls.

This intense "mystical" feel is the exact description of the teacher X. I know.

— Erbao Jian <Freedom Project Manager & Tiehua Village street peddler representative>

Actually, before I did a search on google, I thought Henna was a person. I thought she was the name of that peculiar woman that grabbed my arm and placed it on her thigh, and then used a strange paste to draw beautiful patterns on it.

After googling what it was, I found out that Henna not only can it be used to get intimate with others, it is also an interesting form of art, just like creative painting, photography, and sculpturing etc. In addition, Henna has a rich and interesting classical culture background. Although I am a man, I am still without a doubt attracted by Henna; you are more than welcome to join us if you are interested!

— An <Person in charge of the handcrafted goods & clothing at Old Fashioned Taitung Dormitory>

When I journeyed through India in 2008, I experienced Henna Body Art for the first time. After I came back to Taiwan, I came across images of Henna cards that teacher X. had drawn while browsing the internet. Instantly I was drawn to her classic yet modern and unique Henna patterns. I signed up for her class immediately.

I remember in the first session, X. mentioned that she wanted to lie in a coffin covered with Henna patterns when she leaves this life. I was surprised to hear this at first, but then I felt really moved, because I understood and felt her passion and persistence towards Henna.

It is very lucky when people are able to pursue their dreams in life, and to me, teacher X. is one of them.

— Sharkaren <A student in the Henna beginners' class>

推薦序

Henna，很棒！！

是那纏繞的線條牽引著我的心吧！第一次遇見 Henna，從老師流暢的交織構圖到圖騰能存在一星期的時間，還有淡淡葉子香，都讓人打從心底著迷。

盼到開課的第一天，擠出第一筆顏料是多麼讓人新奇興奮，開始學著構圖之後，更加留神這世界的美麗花紋，在擠壓顏料的過程，是凝神的專注，在完圖的時刻，是快樂的富足。

很感謝可以遇見小美，很感謝有 Henna 相伴。

— March ＜成大解剖所研究生＞

2011 年，臺東加路蘭手創市集，第一次遇見小美老師，也第一次體驗了印度 Henna 彩繪，對於一個喜歡民俗風，嚮往印度風情的我來說，就像老天爺送我的禮物一樣，更讓我受寵若驚的是，小美老師利用短短留在台東的時間，願意撥時間教學 Henna 彩繪，更讓我著迷不已。

學習過程有非常多的樂趣，光學著如何擠出顏料的多和少，就學得指頭快抽筋，構圖也需要一些想像力，用筆畫出圖形，個個沒問題，但每次使用顏料，就老畫得四不像，有時還會腦筋打結到得用電腦向老師求救，小美老師也不吝嗇，當下畫圖附上文字說明，照相傳檔案，來個線上教學，對我來說太感動了。

Henna 或許沒有配色上的問題，但在構圖上是門學問，它可以是柔軟唯美的花圖，也可以是個性剛毅的圖騰，因為是在身上彩繪，所以必須依不同部位給予不同線條；對象不同，也會有不同的視覺感受，絕非三兩下工夫可以繪成的。

Henna 看似簡單，但真的實際拿起顏料管子開始繪畫時，一切都不簡單。線條的穩定度、流暢感，和所有的一角一勾一點的感覺，完全需要純熟的技術，每每看著小美老師為客人繪畫時，那種專注、氣定神閒的樣子，總會讓人投以讚賞的眼光，彩繪完後的客人，就跟我一樣，因自己身上漂亮的 Henna 圖案滿意歡喜，更有人像是上了癮，一畫再畫。

我覺得這是一種緣分，我喜歡 Henna 彩繪，在小小台東，我遇見了小美，我以為我要到印度才有機會畫 Henna，沒想到，小美先讓我小小圓夢，也因如此結交了一位這麼特別的朋友。

有人說，小美像個吉普賽女郎，出現在不同地方為人畫 Henna 彩繪，對我來說，小美是個為夢想活著的女人，她自成一種個人獨特的風格，吸引別人探視她的 Henna 手繪世界！

— CHI ＜ Henna&Chi 品牌工作夥伴＞

Preface

Henna, is the best!!

I think my heart is hooked by the curled and intertwined lines. When I first came in contact with Henna, I saw the preliminary sketch and watched it turn into the Henna pattern in a week's time. Within the week, I emerged in the light smell of leaves, and I became obsessed with everything related to Henna.

Finally, I started my first session; I was so excited to apply the Henna paste for the first time. After I learned how to sketch out a basic design, I started to pay more attention to all the beautiful patterns that could be seen around me. In the application of the Henna paste, during the focused working process and when you complete the design, you will find happiness and satisfaction.

I am so grateful for knowing X., and to be able to have Henna as my companion.

— March<Graduate student at the Institute of Anatomy at National Cheng Kung University>

I first met teacher X. at the Jialulan Handicraft Market in 2011, it was also my first encounter with the Indian Body Art Henna. To someone that is quite fond of ethnic styles and the Indian culture, the encounter was a like present from God. What touched me even more was that although teacher X. was staying in Taitung for a short period of time, she still made time to teach me Henna. I became even more obsessed with this form of art.

The Henna learning process was a lot of fun, just practicing the control of paste amount is enough to make your fingers cramp. Of course sketching up a design also requires some magination. It's always easy when drawing the designs with a pen, but when it comes the time to use the Henna paste, the design often becomes distorted. Sometimes, when the designs are too much and your brain goes on strike, you can even ask for the teacher's help via the Internet. Teacher X. is very generous. She will provide help immediately, by providing instant drawings and explanations. She uses pictures to teach me on-line, I am so glad to have her as my teacher.

Although you do not need to worry about color coordination in Henna, forming the design can be quite a difficult task. Henna patterns can be anything from a soft beautiful floral pattern, to a cool and edgy hard-core pattern.Because Henna uses the skin as the canvas, you can choose different design structures for different body parts; the visual emotions of the pattern drawn also differ from one person to another. The art of Henna is not something one could master in a short period of time.

Henna looks simple, but when you take the applicator in hand and start to draw, you will find that it is not as simple as you may think. The stability and smoothness of the lines, and the corners and angles all require matured skills and techniques.Every time I watch teacher X. draw Henna for her customers, her focus and cool makes her so admirable to others. After the completion, every customer is satisfied with the beautiful Henna design on their bodies, just like I was. In addition, many of them come back for more. It is just like an addiction.

I think it is my destiny that made me fall in love with Henna, and brought me to Taitung to get to know X.. I thought I had to go to India to have a change to draw Henna. Instead X. made my dreams come true right here in Taiwan, and I am blessed to have such a special friend.

Some people say that X. is like a gypsy woman, going from one place to another to draw Henna for people; to me, X. is a woman that lives to fulfill her dream, and her one of a kind style is the reason why others want to come and see for themselves the world of Henna Body Art.

— CHI <Henna&Chi Brand Co-founder>

目錄 CONTENTS

初體驗的第一步──認識 Henna ！

Henna 的由來

　　Henna，又稱 Mehndi，是一項印度傳統手繪藝術，在印度婚禮中，不只女眷身上畫有 Henna，它更是新娘必要的妝飾，在結婚的前一天，資深手繪師以 Mehndi（散沫花）的葉子細磨調合的顏料，在新娘的手腳皮膚上畫滿美麗細緻的圖騰。

　　新娘身上的 Henna 創作，除了美化視覺，也帶有深層意義，以下分出幾項介紹：

　　1. Henna 畫得越多，代表新娘帶給夫家的財富與福氣就越多。

　　2. 從畫上圖騰到顏料完全剝落的時間，新娘不必做家事，這段時間的長短，是婆婆判定新娘能不能帶來富貴與幸福的標準。

　　3. 顏料脫落後，留在皮膚上的顏色越深越好，代表身體越健康。

　　4. Henna 代表娘家對女兒的疼惜，怕女兒出嫁後被夫家虐待，如果 Henna 停留在皮膚上的時間很久，就表示沒有為家事勞累辛苦，沒有受到欺凌。

　　不管是現代還是古代，Henna 就是幸福的代名詞，在印度婚禮中，比較講究的家庭，還會在新娘身上畫出一對象徵新人的人形。

　　介紹完 Henna 的由來，再來欣賞 Henna 的多元圖騰，有些取材自生活，印度因為盛產芒果，因此便有了芒果的圖案與線條，可說是一種生活紀錄；有些則取吉祥意義，像印度的國鳥──孔雀，代表吉祥富貴，是新娘身上一定會看到的圖案。

　　Henna 藝術延伸至今，除了是一項古老的習俗，也變成一種崇尚自然的彩繪藝術，經典圖騰搭配各種創新圖案，提供手繪師自由揮灑的創作空間。

The First Phase of the Henna Experience – Getting to know Henna!

The History of Henna

Henna, also known as Mehndi, is a form of traditional Indian art form used in Indian wedding. Not only do the females attending the wedding have Henna, it is also an indispensable accessory of the bride. On the day prior to the wedding, an experienced artist grinds the leave of the Mehndi to prepare the Henna paste, and draws delicate patterns all over the bride's hands and feet.

The Henna Art work is not only for visual beautification, it also has more profound meanings, which are elaborated below:

1. The more the amount of Henna on the bride, means the more wealth and luck the bride will bring to the groom's family.

2. The bride is forbidden from doing any housework as long as the colour of the bridal Henna remains on her hands, the color duration affects the mother-in-law's perception of whether the bride is able to bring wealth and happiness to the family.

3. The deeper the final color after the paste had wore off, means the healthier the bride is.

4. Henna represents how much the bride will be loved in her husband's family. The fear of the bride being ill-treated by the family she married into may be relieved if Henna is able to remain on her skin for a long period of time. This means that she will not slave over housework, and she is well-treated by her new family.

No matter whether it is the present or the past, Henna stands for happiness just the same. In Indian weddings, a more refined family will choose to draw a humanly designs symbolizing the newly weds.

After getting to know Henna's history, let us take a look at the abundant designs: some were inspired by life, as the mango is India's national fruit, therefore, mangos are massively incorporated into Henna designs as a form of life recording; other designs representing luck and fortune incorporates the peacock, which is India's national bird that stands for fortune and wealth, and indispensible in bridal Henna designs.

Nowadays, the art of Henna is not only an ancient tradition, but also a form of natural art creation. Traditional patterns along with the novel designs provide the Henna artist with the freedom of unlimited creations.

Henna 的材料

　　Henna 的顏料主要是 Mehndi（散沫花，又稱指甲花，產於中亞與印度，本身帶有香氣，也有染髮與香水材料的用途。）的葉子，乾燥磨成粉末後，與水調合而成。

　　每個調製師都有自己的獨家配方，有的添加檸檬汁，有的添加尤加利精油，這些材料大部分是純天然的，本身不傷皮膚，但患有蠶豆症的人不能碰觸 Henna 顏料。

　　最早期的手繪顏料需自行調合，再以竹籤、細木棒之類的工具沾取顏料，於皮膚上繪圖。現在想畫 Henna 比以前方便多了，已有包裝好的便利顏料，只要剪開管口就能使用，如果想體驗手調顏料的樂趣，依然可以買到粉狀原料，以溫的紅茶或咖啡水，再加檸檬汁或糖及 Mehndi 油拌勻，靜置 3-4 小時即可。加入 Mehndi 油可讓顏色持久，彩繪前也可以將 Mehndi 油塗抹在要畫 Henna 的位置，讓美麗的圖騰多停留一段時間。

　　當顏料畫上皮膚後，要有足夠的耐心等待完全乾燥，時間長短需以當時的氣候、空氣濕度而定，越熱越乾燥的氣候，會加速顏料乾燥的速度，之後的圖騰顯色也會因高溫而顯得更加深沉。

　　Henna 停留在皮膚上的時間與顏色，會依氣候、體質、顏料、彩繪位置而不同，但基本上與膚色無關。乾燥後剝去顏料，一般約可維持一星期，但易摩擦或碰水的部位，如手掌，就比較不容易維持。Henna 的顯色，以咖啡色較為多見，如果顏料中加有硃砂或是新磨顏料，色澤會偏紅，加膠顏料的顏色較持久，顏色較鮮豔，但也因為原料的關係，較難擠出。

Henna 顏料（粉末狀）

Henna 管狀顏料

Materials used in Henna

The Henna paste is made by mixing water with dried powdered mehndi leaves (Aka henna or Lawsonia inermis. It grows in central Asia and India, has a unique aroma, and can be found in hair dyes and fragrances).

Every artist has their own unique formula, some add lemon juice, and some add eucalyptus essential oils. Most of the ingredients are 100% natural and harmless to the skin, however, G6PD deficiency patients can not come in contact with the Henna paste.

At the beginning, the Henna paste was 100% homemade, and applied with a thin pick or wooden tool on the skin. The utilization of Henna nowadays is much simpler, with packaged kits of convenient ready-to-use paste that only need to be cut open in order to use. If you want to experience for yourself the fun of preparing the Henna paste from scratch, there are still powdered materials that can be bought in the market. Mix thoroughly into the mehndi oil some warm black tea or diluted coffee, along with lemon juice or sugar, and the paste will be ready to use after 3 to 4 hours. The addition of mehndi oil makes the color last longer, you could also apply the mehndi oil onto the area of skin to work on prior to Henna paste application to make your beautiful Henna designs last longer.

When the paste is applied onto the skin, you must be patient and allow it to dry completely. The drying time depends on the climate and humidity. The drier and hotter the climate is, the easier and quicker the paste will dry, and the darker and more obvious the design will become due to the heat.

The duration of time that Henna is left on the skin is associated with the final coloring, which differs due to the climate, condition of the body, quality of paste, and area of application, but has little to do with skin color. When Henna dries out, excess paste may be picked off, and the design is generally able to last about one week. However, designs on the areas prone to friction or frequently contacts water, such as the palm, are more difficult to maintain. The color of Henna is mostly brown in color, if the paste contains cinnabar or freshly ground pigments, the coloring will become redder; the addition of PVC pigments are more long-lasting and brighter in comparison, however, harder to apply due to its' texture.

powdered ingredients

Tube colors

其他國家的 Henna 藝術

　　除了印度，巴基斯坦、伊朗、摩洛哥、阿拉伯等國家也有 Henna 手繪藝術，雖然因時間太過久遠而無法考證起緣，但意義與創作精神是相同的，都為了美麗而存在。

　　隨著不同國家，繪製出的圖騰也不盡相同，各有各的風格，以非洲國家為例，圖騰以幾何線條為主，地處北非的摩洛哥就呈現了該種風格；越靠近中亞地區，線條越顯得柔美，帶有纏繞感。

　　除了上述的差異之外，顏料外形也會不一樣，像馬來西亞使用的顏料，就有軟管牙膏狀、一管一管的筆狀等，顏色也會有深淺的差別。

Henna 的多元運用

　　Henna 不只可做人體彩繪，還能有更多元的運用，以不同的工具與手技，做出完全不一樣的作品。例如，進行布類彩繪時，可利用壓克力顏料，畫出衣服或包包上的圖案；不上漆木材則可使用一般的 Henna 顏料，或壓克力顏料。

體驗 Henna 的地點

　　想體驗手繪師在自己身上進行 Henna 的感覺，又沒辦法跑去印度嗎？別擔心！在臺灣也可以體驗喔～只要上網搜尋一下創意市集，就可以找到手繪 Henna 的攤位囉！

The differences of Henna in different countries

Aside from India, Pakistan, Iran, Morocco, and Arab, other regions also have art forms based with Henna, although they are so ancient to the extent that their origins are lost in history, the meaning and spirit behind these creations are similar, they were all created in the name of beauty.

Different countries develop different designs, each with their own unique style. African designs for example, mainly consist of symmetrical lines, such as the Moroccan styles in Northern Africa; designs from the central Asian region on the other hand are consisted of softer lines and intertwined features.

Aside from the aforementioned differences, the appearance of paste in different regions is also different from one another. For example, the Malaysian paste are packaged in toothpaste-like tubes, or in pen shaped tubes. The darkness of coloring varies as well.

The multiple uses of Henna

Henna can be used in Body Art, but also in a variety of occasions with different tools and techniques to create a completely different form of art. Acrylic pigments can be used on fabric to create garments or handbags. The common Henna paste or acrylic pigments can be used for art on untreated wood.

The place to go to experience Henna

What can you do if you want to experience for yourself the application of Henna on your skin, and you can't go to India? Don't worry! You can have the Henna experience right here in Taiwan. Just search for creativity markets online and you'll know where to find the Henna booth.

Henna 材料介紹　Introduction to Henna application tools

管狀顏料 Tube colors

以粉狀原料為主的調合顏料，用手指擠壓的方式進行 Henna 手繪。剪開顏料管口後，不使用時須以大頭針塞住管口，以防空氣進入管內，造成顏料乾燥無法使用。未開封或未使用完的顏料，以室溫保存即可。

Mainly formulated with powdered ingredients, and applied by squeezing the tube with fingers. After cutting open the tip of the tube, the opening is plugged with a pin to avoid the contact of air that causes the coloring to dry up. Unused or unfinished tubes may be stored at room temperature.

Mehndi 油 Mehndi oil

用於調和 Henna 顏料，可使顏色更持久，亦可在身體彩繪前，取適量塗抹於作畫部位，延長 Henna 顯色的時間。

Mehndi oil is used in the formulation of Henna paste, and can be used to make the colors last longer. It can also be applied onto the area prior to Henna paste application, for a long-lasting Henna design.

噴霧式水瓶 Spray bottle

在 Henna 完全乾燥前，可在圖騰上噴水，延緩顏料剝落時間，使顯色與輪廓更明顯。

Water can be sprayed on the Henna design before it is completely dry in order to postpone the peeling of the paste and create a design with more defined lines and brighter color.

PART 1
基礎技巧

想學會Henna的精湛手技，基本功一定要做得扎實，
簡單的線條與圖案，都包含了大學問！

If you want to become a master in Henna Body Art,
you must lay the groundwork.
The lines and designs may seem simple,
but there's much more than meets the eye!

握筆
Holding the applicator

以圖片的手指位置握筆,各部位的施力須均勻,管口要直。擠出顏料前,管口稍微懸空,避免顏料堆積,或運筆時產生分岔線條。

Hold the applicator as shown with evenly distributed pressures. Before squeezing out the paste, the tip must be lifted up a few millimeters from the skin or surface of application to avoid paste buildup and split lines.

如果進行 Henna 的人體彩繪,運筆時沿著身體曲線調整位置,不須一直維持直立狀態。顏料即將用盡時,握筆位置可以下方為重心,動作以輕鬆施力,能均勻擠出顏料為準。

If the Henna paste is used for doing body art, the position of the applicator must be adjusted according to the natural curves of the body part. The applicator does not have to maintain an upright angle. When the paste is running out, you may hold the applicator closer to the tip. Movements should focus on the ease of application, and the even distribution of the Henna paste.

直線
延伸應用:
斜線、平行線、交叉線、三角形、菱形
Straight lines
Extended applications:
Diagonal lines, parallel lines, crossed lines, triangles, and diamonds.

1

先擠出一滴顏料。
First squeeze out the Henna paste.

2
以均勻的力道將顏料往右拉直。
Then, use even pressure to extend the line to the right.

圖型繪製
Draw

Fin
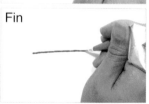
將直線拉至所需的長度即可。
Extend the line as needed and complete the line.

弧線
Curved lines

延伸應用：
圓圈、橢圓、螺旋、藤蔓、延伸型水滴、變形蟲輪廓
Extended applications：
Circles, ovals, spirals, vines, elongated drop-shapes,
and amoeba-shaped outlines.

圖型繪製
Draw

1
先擠出一滴顏料。
First squeeze out the Henna paste.

2
以均勻的力道將顏料往右上拉出弧度。
Then, use even pressure and extend the line curved upward to the right.

Fin
將弧線拉至所需的高度，再以反方向的弧度往下收起即可。
Extend the stroke upward to the height needed, and then curve downward to finish the curve.

曲線
Waves

延伸應用：
花邊
Extended applications：
Waved borders.

圖型繪製
Draw

1
先擠出一滴顏料。
First squeeze out the Henna paste.

2
以均勻的力道將顏料往右邊拉出上下弧度，線條粗細須一致。
Then, use even pressure to make waves extended to the right, the thickness of the line should be consistent.

Fin
將曲線拉至所需的長度即可。
Extend the waves as needed.

粗細變化—粗細線條
Thickness control – Line thickness

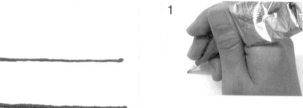

1

先擠出第一滴顏料。
First squeeze out the Henna paste.

2

以放輕、均勻的力道將顏料往右拉出線條。
Use a light and even pressure to extend the paste into a line.

圖型繪製
Draw

I  （細）

3

將線條拉至所需的長度後，完成細線。
Extend the thin line as needed.

II  （細）（粗）

4

在第一條線的下方擠出第二滴顏料。
Squeeze out the Henna paste below the just finished line.

5

以加重、均勻的力道將顏料往右拉出線條。
Use a heavy and even pressure to extend the paste into a line.

Fin

將線條拉至所需的長度後，完成粗線。
Extend the thick line as needed.

粗細變化— 粗細漸層
Thickness control – From thin to thick, from thick to thin

圖型繪製
Draw

I

（細）　　　　　　（粗）

II

（粗）　　　　　　（細）

1

先擠出第一滴顏料。
First squeeze out the Henna paste.

2

以均匀的力道將顏料往右拉出線條。
Then, use even pressure to extend the line to the right.

3

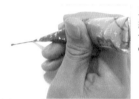

將線條拉至想加粗的長度前端，開始慢慢加重力道，擠出較多的顏料，再拉至所需的長度。
Extend the line to where it should start to become thicker, and gradually apply heavier pressure to squeeze out more paste, and then extend the line as needed.

4

在第一條線的下方擠出第二滴顏料。
Squeeze out the Henna paste below the just completed line.

5

以均匀的力道將顏料往右拉出線條。
Then, use even pressure to extend the line to the right.

Fin

將線條拉至想收細的長度前端，開始慢慢放輕力道，擠出較少的顏料，再拉至所需的長度即可。
Extend the line to where it should start to become thinner, and gradually apply lighter pressure to squeeze out more paste, and then extend the line as needed.

作品急救法—斷線修補
Mistake correction – Fixing a broken line

剛開始手繪 Henna 時，會因為無法控制力道而使線條中斷，產生尷尬的空白。以下教大家如何急救斷線的作品，只要三個步驟喔！

Henna beginners often encounter breaks in their lines due to pressure control inabilities, presenting an awkward blank in the middle. The steps taken to fix a broken line are further elaborated below.

註：進行修補時，切記不可一次擠出大量顏料，才不會影響線條原本的粗細。
Tip： When fixing a line, remember not to squeeze out too much paste at one time in order to maintain the consistency of the line thickness.

1 在斷線的空白處輕輕擠出一滴顏料。
Squeeze out the Henna paste gently at the blank site.

2 將顏料輕輕往右拉，補滿空白處。
Extend the paste to the right to fill in the blank.

Fin 再將顏料輕輕往左拉，稍微調整線條弧度與粗細即可。
Then go backwards to the left to adjust the line thickness and curve.

作品急救法—曲線修補
Mistake correction – Fixing a curve

手繪 Henna 時，會因為力道不均或施力方向偏離，而使作品產生醜醜的曲線。以下教大家如何急救出糗的作品，只要三個步驟喔！

When painting Henna, sometimes uneven pressure or a direction slip may result in unsightly curves. The steps taken to fix a mistake are further elaborated below.

註：進行修補時，切記不可一次擠出大量顏料，才不會影響線條原本的粗細。
Tip：When fixing a line, remember not to squeeze out too much paste at one time in order to maintain the consistency of the line thickness.

1

在曲線上下輕輕擠出顏料。
Squeeze out the Henna paste gently onto the curve that needs to be fixed.

2

以顏料修平曲線，但弧度須與作品一致。
Use the paste to smooth out the curve line, at the same time maintaining the original curviness.

Fin

最後再以顏料稍微調整線條的粗細即可。
Lastly, just use the paste to adjust the thickness of the line.

點點系列—小水滴
Drops – Water drops

1　先擠出一滴顏料。
First squeeze out a small drop of Henna paste.

2　將顏料輕輕往後拉出弧度。
Then, lightly extend the paste to create a small radian.

圖型繪製
Draw

Fin　拉至所需長度後，將筆尖輕輕往上提起收尾。
Extend the radian as needed, and lift the applicator tip to finish.

點點系列—火花
Drops – Firework pattern

步驟説明
Step by step

1　先擠出一滴顏料，畫出第一滴水滴。
First squeeze out a small drop of Henna paste, and draw the first water drop.

2　依序往右畫出水滴，排成半圓弧形，須注意大小、間隔，與排列的弧度。
Draw water drops by clockwise direction to form the right portion. Aware of the size, the space in between, and the alignment curviness of the water drops.

圖型繪製
Draw

Fin　再依序往左畫出水滴，注意大小、間隔，與排列的弧度，排成圓形即可。
Lastly, draw water drops counter-clockwise direction to form the left portion and complete a full circle of water drops to the firework pattern. Aware of the size, the space in between, and the alignment curviness of the water drops.

三角形系列—三角形
Triangles – Simple triangle

圖型繪製
Draw

1

先擠出一滴顏料，往左下拉出直線。

First squeeze out the Henna paste and extend to the lower left to form a straight line.

2

將第一條直線拉至尾端，再往右拉出水平線。

Then, start from the bottom end of the previous line and extend a horizontal line to the right.

Fin

將水平線拉至尾端，再往左上拉出直線，與第一條直線密合即可。

Lastly, start from the right end of the horizontal line and draw a straight line that connects it with the top end of the first line to finish the triangle.

三角形系列—鋸齒
Triangles –Jagged line

圖型繪製
Draw

步驟說明
Step by step

1

先擠出一滴顏料，往右拉出直線。

First squeeze out the Henna paste and extend to the right to form a straight line.

2

往下畫出 V 字形，呈現第一個三角形。

Draw a V-shape to create the first triangle.

Fin

依序往直線右方畫出 V 字形，須注意大小與間隔，畫至尾端後即可完成。

Continue to draw V-shapes to the right under the horizontal line. Remember to mind the size and space between triangles. Carry on until you've reached the end of the line.

圓形系列—圓圈
Circles – Simple circle

圖型繪製
Draw

1 先擠出一滴顏料。
First squeeze out the Henna paste.

2 往右下畫出圓弧線。
Draw a round curve in a downward right direction.

Fin 再往左上畫出圓弧線，與步驟 2 的線條起點相接後即可完成。
Lastly, draw a round curve in an upward left direction, and connect the line to form a circle.

圓形系列—同心圓
Circles – Concentric circle

圖型繪製
Draw

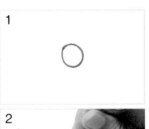

1 先擠出一滴顏料，畫出第一個圓圈。
First squeeze out the Henna paste and draw the first circle.

2 沿著第一個圓圈內側，由下往上畫圓弧線。
Then, draw a round curve from bottom to top inside the first circle.

Fin 圓弧線頭尾相接後，即完成同心圓圖案。
Lastly, connect the line to finish the inner circle and complete the concentric circle.

螺旋系列— 左旋
Spirals – Clockwise spiral

延伸應用：
藤蔓、變形蟲輪廓
Extended applications：
    vines, amoeba-shaped outlines.

圖型繪製
Draw

步驟説明
Step by step

1　先擠出一滴顏料，往左下畫出圓弧線。

First squeeze out the Henna paste and draw a round curve in the downward left direction.

2　將圓弧線往右上拉成圓圈，再往內側收，線條之間不密合。

Then, extend the round curve in the upward right direction and slightly to the center, avoiding connection of the line.

Fin　將步驟 2 的線條再往左上拉成圓圈，線條之間不密合。往內重複步驟 2-3，畫出需要的大小即可。

Extend the line in step 2 in the upward left direction, also remembering to slant slightly to the center, making sure the lines do not meet. Repeat steps 2 to 3 and draw inwardly to for the spiral as needed.

螺旋系列— 右旋
Spirals – Counterclockwise spiral

圖型繪製
Draw

步驟説明
Step by step

1　先擠出一滴顏料，往右下畫出圓弧線。

First squeeze out the Henna paste and draw a round curve in the downward right direction.

2　將圓弧線往左上拉成圓圈，再往內側收，線條之間不密合。

Then, extend the round curve in the upward left direction and slightly to the center, avoiding connection of the line.

Fin　將步驟 2 的線條再往右上拉成圓圈，線條之間不密合。往內重複步驟 2-3，畫出需要的大小即可。

Extend the line in step 2 in the upward right direction, also remembering to slant slightly to the center, making sure the lines do not meet. Repeat steps 2 to 3 and draw inwardly to for the spiral as needed.

藤蔓系列—左旋藤蔓
Vines – Clockwise spiral vine

1

先畫出所需大小的螺旋，再將最後一條線往下拉。

First draw a clockwise outward spiral as needed, and then extend the tail downward.

Fin

將最後一條線往右下畫出弧度後，提筆收尾即可。

Complete the vine by curving in a downward right direction and lift the applicator tip to finish.

圖型繪製
Draw

I

II

藤蔓系列—右旋藤蔓
Vines – Counterclockwise spiral vine

1

先畫出所需大小的螺旋，再將最後一條線拉至頂端。

First draw a counterclockwise outward spiral as needed, and then extend the curve to the top of the spiral.

Fin

將最後一條線往左下畫出弧度後，提筆收尾即可。

Complete the vine by extending the line in a downward left direction into a slight curve and lift the applicator tip to finish.

圖型繪製
Draw

I

II

基本構圖 1 — 延伸型水滴
Basic constructions – Extended drop shape

步驟説明
Step by step

圖型繪製
Draw

I

II

1

先擠出一滴顏料，往右上拉出弧線。

First squeeze out a small drop of Henna paste, and draw a curve in the upper right direction.

2

將弧線拉至所需長度後，往下彎成圓弧線。

Extend the curve as needed, and then curve downwards to form a rounded curve.

3

將圓弧線往左下收起，靠近原本的弧線。

Extend the rounded curve line closer and closer to the original curve.

Fin

最後將圓弧線尾端收至弧線中段即可。

Lastly, join the lines at the center of the first curve to finish.

基本構圖 2 — 變形蟲輪廓
Basic construction 2 – Amoeba outline

圖型繪製
Draw

I

II

III

步驟說明
Step by step

1

先擠出一滴顏料，往左上畫出螺旋。

First squeeze out a small drop of Henna paste, and draw a spiral.

2

將螺旋往下拉出弧線後，再往右拉出圓弧線。

Extend the end of the spiral downwards into a curved line, and then extend the curve to the right to form a rounded curve.

3

將圓弧線往左上收成水滴形。

Extend the rounded curve line to the upper left direction to form a drop shape.

Fin

最後將水滴尾端收至螺旋頂端即可。

Lastly, join the drop shape tail at the top of the spiral to finish.

基本構圖 3 — 葉子
Basic construction 3 – The leaf design

圖型繪製
Draw

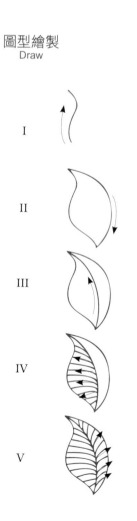

I

II

III

IV

V

1

先擠出一滴顏料，往左上畫出波浪狀曲線。

First squeeze out a small drop of Henna paste, and draw a waved curve in the upper left direction.

2

往下拉出弧線連接到波浪曲線頂點，完成葉子外形。

Extend the curve as needed and draw a curve downward to form a leaf shape. Then draw a line curved line starting at the bottom joining ends at the top to form the leaf shape.

3

在曲線頂點往上拉出弧線，完成中心葉脈。

Extend an upward curve starting from the base of the leaf to create the main leaf vein.

4

在葉片尾端擠出顏料，由下往上依序畫出左側葉脈，線條須略呈放射狀。

Squeeze some Henna paste at the base of the leaf shape, and draw the leaf veins on the left starting from the bottom and moving upward, the leaf veins should be in slightly radial directions.

5

將左側葉脈畫滿，須注意間隔與線條。

Complete the leaf veins on the left, take note of the spaces between each vein.

6

重複步驟 4，由下往上依序畫出右側葉脈。

Repeat step 4, and draw the leaf veins on the right starting from the bottom and moving upward.

Fin

最後將右側葉脈畫滿，即完成葉片圖案。

Lastly, complete the right leaf veins to finish the leaf design.

基本構圖 4 — 花朵系列 刮線花蕊
Basic construction 4 – Flowers: the flicked pistils

步驟説明
Step by step

1 先擠出一滴顏料，畫出尾端密合的螺旋。
First squeeze out a small drop of Henna paste, and draw a spiral that closes at the end.

圖型繪製
Draw

2 以螺旋尾端為起點，依序往下畫出圓形花蕊，須注意大小與間隔。
Using the closed end of the spiral as the starting point, draw rounded pistils to circle the spiral, take note of the size and spacing of pistils.

I

3 圓形花蕊環繞螺旋一圈後，完成花心圖案。
After drawing the full circle of pistils, the core of the flower is completed.

II

4 以圓形花蕊為起點，往左畫出第一片心形花瓣。
Using the rounded pistils as the starting point, draw a heart-shaped petal to the left.

III

5 以第一片花瓣尾端為起點，往下畫出第二片心形花瓣，花瓣之間的線條須密合。
Using the end of the first petal as the starting point, draw the second heart-shaped petal, the petals should be aligned tightly without gaps.

IV

步驟説明
Step by step

6

重複步驟 5，依序由下往上畫出所有的花瓣。

Repeat step 5 and finish all the petals accordingly.

7

在圓形花蕊的頂點擠出少許顏料。

Squeeze a small amount of Henna paste at the top of the rounded pistils.

8

以往外輕刮的方式刮出步驟 7 的線條，即成刮線花蕊。

With a slight outward flicking gesture, use the Henna paste in step to create the light lines as shown in the diagram.

9

依序刮出花蕊時，須注意線條與長度。

Finish all the pistils in the same manner described in step 8, take note of the length of each flick.

Fin

刮出全部的花蕊後，外觀呈放射狀，即可完成花朵圖案。

When the flicked pistils are completed, they will radial appearance, thus completes the flower design.

基本構圖 4 — 花朵系列 拉線花蕊
Basic construction 4 – Flowers: the drawn pistils

圖型繪製
Draw

I

II

III

IV

步驟説明
Step by step

1

先擠出一滴顏料，畫出尾端密合的螺旋。

First squeeze out a small drop of Henna paste, and draw a spiral that closes at the end.

2

以螺旋尾端為起點，依序往下畫出圓形花蕊，須注意大小與間隔。

Using the closed end of the spiral as the starting point, draw rounded stamens to circle the spiral, take note of the size and spacing of stamens.

3

圓形花蕊環繞螺旋一圈後，完成花心圖案。

After drawing the full circle of stamens, the core of the flower is completed.

4

以圓形花蕊為起點，往左畫出第一片心形花瓣。

Using the rounded stamens as the starting point, draw a heart-shaped petal to the left.

5

以第一片花瓣尾端為起點，往下畫出第二片心形花瓣，花瓣之間的線條須密合。

Using the end of the first petal as the starting point, draw the second heart-shaped petal, the petals should be aligned tightly without gaps.

步驟說明
Step by step

6

重複步驟 5，依序由下往上畫出所有的花瓣。

Repeat step 5 and finish all the petals accordingly.

7

在圓形花蕊的頂點擠出少許顏料。

Squeeze a small amount of Henna paste at the top of the rounded pistils.

8

以往外輕拉的方式畫出步驟 7 的線條，即成拉線花蕊。

With a gentle outward stroke, use the Henna paste in step to create the lines as shown in the diagram.

9

依序拉出花蕊時，須注意位置與長度。

Finish all the pistils in the same manner described in step 8, take notice of the length of each stroke.

Fin

拉出全部的花蕊後，外觀呈放射狀，即可完成花朵圖案。

When the pistil strokes are completed, they will radial appearance, thus completing the flower design.

PART 2

初級構圖

學會基本圖案與線條，可以來嘗試一些構圖的設計，
先從初級構圖練習，慢慢加深自己的功力！

After you've become acquainted with simple patterns and lines,
you may try some compositional designs. Starting with the basics,
and practicing to become better and better!

橋形紋

The bridge-like design

弧形構圖1
Curved constructions 1

步驟説明

1

先擠出一滴顏料，往右上
畫出弧線。

2

以平行的方式，在第一條
弧線下畫出第二條弧線。

3

在兩條弧線之間，依序由
上往下畫出直線，須注意
線條與間隔。

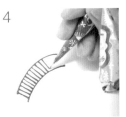

4

畫出直線時，也須配合弧
線弧度調整排列位置。

5

最後將直線畫至所需長度，
即可完成圖案。

Step by step

1 First squeeze out a small drop of Henna paste, and draw a curve in the upper right direction.
2 Then, draw a parallel curve underneath the first curve.
3 Next, draw vertical lines in the up-bottom direction along the curve, taking notice of spaces between each line.
4 Remember to follow the curve when aligning the vertical lines.
5 Lastly, finish the vertical lines as needed to complete the design.

螺旋飾邊圖
Spiral boarder design
弧形構圖2
Curved constructions 2

步驟說明

1

先擠出一滴顏料，往右上畫出弧線。

2

以平行的方式，在第一條弧線下畫出第二條弧線。

3

在兩條弧線之間，畫出第一個螺旋。

4

沿著第一個螺旋右側，畫出第一條弧線。

5

在步驟4的弧線右側，畫出第二條弧線。

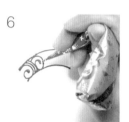

6

以第二條弧線頂端為起點，往下畫出第二個螺旋。

7

沿著第二個螺旋右側，畫出第一、二條弧線。

8

重複步驟3-7，畫至所需長度後，即完成圖案。

Step by step

1. First squeeze out a small drop of Henna paste, and draw a curve in the upper right direction.
2. Then, draw a parallel curve underneath the first curve.
3. Next, draw a spiral between the curves.
4. Draw a curved line along the right side of the spiral as shown in the diagram.
5. Then, draw another curved line along the curved line in step 4 as shown in the diagram.
6. Use the top of the curved line in step 5 as the starting point, draw a second spiral.
7. Then, draw 2 parallel curved lines along the right side of the second spiral.
8. Repeat steps 3 through 7 as needed to complete the design.

編織 線條
Weaved pattern

三角形構圖1
Triangle construction 1

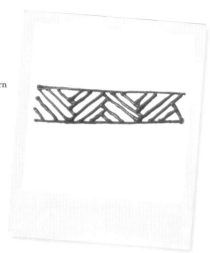

步驟説明

1

先擠出顏料，畫出兩條上下平行的直線。

2

以上方直線為起點，在兩條直線間畫出兩個 V 字，將間隔分成三個三角形。

3

沿著第一個倒三角形內側，畫出第一條內側斜線。

4

以第一條斜線尾端為起點，畫出第二條斜線。須注意間隔，線與線之間不密合。

5

沿著第一條斜線內側，畫出第三條斜線。

6

以第三條斜線尾端為起點，畫出第四條斜線，完成第一個倒三角形圖案。

7

沿著第一個倒三角形左側，畫出第一條外側斜線。

8

沿著第一條外側斜線，往外畫出第二條外側斜線。

Step by step

1 First squeeze out a small drop of Henna paste, and draw 2 horizontal lines that are parallel from each other.

2 Starting from the upper line, draw 2 v-shapes between the 2 lines creating 3 triangles.

3 Draw a 1st parallel line inner the left hypotenuse of the first reverse triangle as shown in the diagram.

4 Then, starting at the bottom of the 1st parallel line, draw a 2nd line that is parallel to the inner side of the right hypotenuse of the first reverse triangle as shown in the diagram. Remember to mind the spaces in between each line.

5 Draw the 3rd line parallel to the 1st line as shown in the diagram.

6 Starting at the bottom of the 3rd line, draw a 4th line parallel to the 2nd line as shown in the diagram and complete the first reverse triangular pattern.

7 Draw a line parallel to the outer side of the first reverse triangl's left hypotenuse as shown in the diagram.

8 Draw the 2nd outer line parallel to the 1st outer line.

9

沿著第二個正三角形內側，畫出第一條內側斜線。

10

以第一條斜線頂端為起點，畫出第二條斜線。須注意間隔，線與線之間不密合。

11

沿著第一條斜線內側，畫出第三條斜線。

12

以第三條斜線頂端為起點，畫出第四條斜線，完成第二個正三角形圖案。

13

沿著第三個倒三角形內側，畫出第一條內側斜線。

14

以第一條斜線尾端為起點，畫出第二條斜線。須注意間隔，線與線之間不密合。

15

沿著第一條斜線內側，畫出第三條斜線。

16

以第三條斜線尾端為起點，畫出第四條斜線，完成第三個倒三角形圖案。

17

重複步驟 2-16，畫至所需長度即可。

9 Draw a 1st parallel line inner the left hypotenuse of the second triangle as shown in the diagram.
10 Then, starting at the top of the 1st parallel line, draw a 2nd line that is parallel to the inner side of the right hypotenuse of the second triangle as shown in the diagram. Remember to mind the spaces in between each line.
11 Draw the 3rd line parallel to the 1st line as shown in the diagram.
12 Starting at the top of the 3rd line, draw a 4th line parallel to the 2nd line as shown in the diagram and complete the second triangular pattern.
13 Draw a 1st parallel line inner the left hypotenuse of the third reverse triangle as shown in the diagram.
14 Then, starting at the bottom of the 1st parallel line, draw a 2nd line that is parallel to the right side of the third reverse triangle as shown in the diagram. Remember to mind the spaces in between each line.
15 Draw the 3rd line parallel to the 1st line as shown in the diagram.
16 Starting at the bottom of the 3rd line, draw a 4th line parallel to the 2nd line as shown in the diagram complete the third reverse triangular pattern.
17 Repeat steps 2 through 16 as needed to complete the design.

山形紋

The mountain-like pattern

三角形構圖2
Triangle construction 2

步驟說明

1

先擠出顏料，畫出兩條上下平行的直線。

2

以上方直線為起點，在兩條直線間畫出兩個 V 字，將間隔分成三個三角形。

3

沿著第一個倒三角形左側，畫出第一條外側斜線。

4

沿著第一條外側斜線，往外畫出第二條外側斜線。

5

以第二條外側斜線頂點為起點，往下畫一條直線，成一直角三角形。

6

以顏料補滿步驟 5 的直角三角形。

7

沿著第一個倒三角形內側，畫出第一個 V 字。

8

沿著第一個 V 字形內側，畫出第二個 V 字，此時 V 字與直線夾成倒三角形。

Step by step

1　First squeeze out a small drop of Henna paste, and draw 2 horizontal lines that are parallel from each other.
2　Starting from the upper line, draw 2 V-shapes between the 2 lines creating 3 triangles.
3　Draw a line parallel to the outer side of the first reverse triangle's left hypotenuse as shown in the diagram.
4　Draw the 2nd outer line parallel to the 1st outer line.
5　Then, starting from the top of the 2nd outer line, draw a vertical line to from a right triangle.
6　Use the Henna paste to fill in the right triangle created in step 5.
7　Inner to the first reverse triangle, draw a 1st V-shape with both sides of the v parallel to the hypotenuses of the triangle.
8　Then, draw a 2nd V-shape inner to the 1st V-shape. The 2nd V-shape forms a reverse triangle with the upper horizontal line.

9

以顏料由外而內補滿步驟 8 的倒三角形。

10

補滿圖案時，顏料須均勻無空隙。

11

沿著第二個正三角形內側，畫出第一個倒 V 字。

12

沿著第一個倒 V 字形內側，畫出第二個倒 V 字，此時倒 V 字與直線夾成正三角形。

13

以顏料補滿步驟 12 的正三角形。

14

沿著第三個倒三角形內側，畫出第一個 V 字。

15

沿著第一個 V 字形內側，畫出第二個 V 字，此時 V 字與直線夾成倒三角形。

16

以顏料補滿步驟 15 的倒三角形。

17

重複步驟 7-16，畫至所需長度即可。

9 In an inward direction, use the Henna paste to fill in the reverse triangle created in step 8.

10 The Henna paste should be distributed evenly without leaving gaps when filling in the patterns.

11 Inner to the second triangle, draw a 1st reversed V-shape with both sides of the v parallel to the hypotenuses of the triangle.

12 Then, draw a 2nd reversed V-shape inner to the 1st one. The 2nd reversed V-shape forms a right triangle with the lower horizontal line.

13 Use the Henna paste to fill in the right triangle created in step 12.

14 Inner to the third reverse triangle, draw a 1st V-shape with both sides of the v parallel to the hypotenuses of the triangle.

15 Then, draw a 2nd V-shape inner to the 1st V-shape. The 2nd V-shape forms a reverse triangle with the upper horizontal line.

16 Use the Henna paste to fill in the right triangle created in step 15.

17 Repeat steps 7 through 16 as needed to complete the design.

延伸型螺旋 Extended spirals

螺旋構圖1
Spiral construction 1

步驟説明

1	2	3

先擠出一滴顏料，往下畫出第一個螺旋。

以第一個螺旋下方為起點，往下畫出第二個螺旋。

以第二個螺旋下方為起點，重複步驟 2，畫至所需長度即可。

Step by step

1 First squeeze out the Henna paste and draw a downward clockwise spiral.
2 Using the bottom of the clockwise spiral as the starting point, draw a downward counterclockwise spiral.
3 Using the bottom of the counterclockwise spiral as the starting point, repeat the 2 steps above as needed to complete the design.

延伸型藤蔓 Extended vines

螺旋構圖2
Spiral construction 2

步驟說明

1

先擠出第一滴顏料,畫出第一個右旋藤蔓。

2

以第一個右旋藤蔓中段為起點,往左上拉出弧線。

3

將第二條弧線往內畫成左旋藤蔓,完成底部。

4
將第三條弧線往下畫成螺旋,以左旋藤蔓右側為起點,往左上拉出第三條弧線,拉至中段時,再往右拉出明顯弧度。

5

將第三條弧線往下畫成螺旋,完成右旋藤蔓。

6

以左旋藤蔓頂端為起點,往右上拉出第四條弧線。

7

第四條弧線拉至頂端,再往左拉出圓弧線。

8

最後將第四條圓弧線往內畫成左旋,完成第二條左旋藤蔓,即可完成延伸型藤蔓。

Step by step

1 First squeeze out the Henna paste and draw a counterclockwise spiral vine.
2 Starting from the midpoint of the counterclockwise spiral, draw a curved line in the upper left direction.
3 Extend the left curve inward into a clockwise spiral vine, completing the base.
4 Starting from the junction of the 2 spirals, draw a 3rd curve in the upper left direction. Then extend the curve and make an obvious right turn.
5 Extend the turn of the 3rd curve into a spiral to complete the counterclockwise spiral vine.
6 Starting from the top of the clockwise spiral vine, draw a 4th curve in the upper right direction.
7 Extend the 4th curve to the top, and make a left turn.
8 Lastly, extend the 4th curve inward into a clockwise spiral to finish the 2nd clockwise spiral vine and complete the design.

蝶形飾紋
The butterfly-like pattern
螺旋構圖3
Spiral construction 3

步驟説明

1

先擠出一滴顏料，往下畫出第一個左旋。

2

將左旋尾端往下拉出弧線。

3

以弧線尾端為起點，往上畫出波浪曲線。

4

將波浪曲線尾端往上拉起，收入左旋內側，完成輪廓。

5

在輪廓的尾端夾角，畫出第一個螺旋。

6

沿著螺旋上方，依序往上畫出弧線。

Step by step

1 First squeeze out the Henna paste and draw a downward clockwise spiral.
2 Extend the tail of the clockwise spiral into a downward curve.
3 Using the end of the curve as the starting point, draw an upward m-contourd wave as shown in the diagram.
4 Extend the tail of the wave upward and join the tail into the inner side of the clockwise spiral.
5 At the bottom corner of the just drawn contour, draw a spiral.
6 Draw parallel curved lines above the spiral until reaching the mid-point of the contour.

7	8	9	10

在輪廓的頂端夾角，畫出另一個螺旋。	沿著步驟 7 的螺旋下方，依序往下畫出弧線。	沿著步驟 7 的螺旋上方，依序往上畫出弧線。	以輪廓的尾端為起點，往上畫出第二個右旋。

11	12	13	14

以輪廓的尾端與第二個右旋的交界為起點，往右上拉出弧線。	將步驟 11 的弧線頂端往下拉，畫出第三個螺旋。	以輪廓右側中點為起點，往上拉出弧線。	最後將步驟 13 的弧線頂端往下拉，畫出第四個螺旋，即可完成作品。

7   Then, draw another spiral at the mid-point of the contour.

8   Draw parallel curved lines below the 7rd spiral.

9   Draw parallel curved lines above the 7rd spiral.

10   Starting from the bottom tip of the previously drawn contour, draw the second conterclockwise spiral as show in the diagram.

11   Starting from the junction of the bottom tip of the contour and the second conterclockwise spiral, draw a curved line in the upper right direction.

12   Then extend the curved line in step 11 downward to create the third spiral.

13   Starting from the midpoint of the curve created in step 2, draw a curve in the upward direction.

14   Lastly, extend the curved line in step 13 downward to create the fourth spiral to complete the design.

羽狀飾紋
螺旋構圖4
The feather-like pattern | Spiral construction 4

步驟説明

1

先擠出一滴顏料,往左下拉出弧線。

2

將弧線尾端往上拉,畫出第一個右旋。

3

以弧線頂端為起點,往下畫出波浪曲線,曲線尾端往右旋內側收起,完成輪廓。

4

以輪廓的頂端為起點,往左畫出右旋。

5

以輪廓的頂端為起點,在步驟4的螺旋下方往左拉出弧線,尾端收成右旋,收至尾端時,多擠一滴顏料,做出水滴效果。

6

以輪廓的頂端為起點,往右畫出左旋。

Step by step

1 First squeeze out the Henna paste and draw a downward curve to the left.

2 Extend the tail of the curve upward and create a counterclockwise spiral.

3 Using the top end of the curve as the starting point, draw a downward m-shaped wave as shown in the diagram. Extend the tail of the wave downward and join the tail into the inner side of the counterclockwise spiral.

4 Using the top tip of the contour as the starting point, draw a counterclockwise spiral to the left.

5 Using the same starting point, draw a curve extending under the spiral in step 4, and then extend the curve to create another counterclockwise spiral. Squeeze out a small excess amount of Henna paste at the end of the spiral to from a drop-like contour.

6 Start at the top of the contour, and draw a clockwise spiral to the right.

7

以輪廓的頂端為起點，在步驟 6 的螺旋下方往下拉出弧線，收至尾端時，多擠一滴顏料，做出水滴效果。

8

以輪廓的頂端為起點，在步驟 5 的線條下方往下拉出弧線，尾端收成左旋。螺旋收至尾端時，多擠一滴顏料，做出水滴效果。

9

在輪廓內側的前段，畫出一個左旋。

10

沿著步驟 9 的螺旋，依序往上畫出弧線。

11

沿著步驟 9 的螺旋，依序往下畫出弧線。

12

在輪廓內側的後段，畫出第二個螺旋。

13

沿著步驟 12 的螺旋，依序往下畫出弧線。

14

最後沿著步驟 12 的螺旋，依序往上畫出弧線，即可完成圖案。

7 Using the same starting point, draw a curve extending under the spiral in step 6, and then extend the curve and squeeze out a small excess amount of Henna paste at the end of the curve to from a drop-like end.

8 Using the same starting point again, draw a curve extending under the spiral in step 5, and then extend the curve to form a clockwise spiral. Squeeze out a small excess amount of Henna paste at the end of the spiral to from a drop-like end.

9 Then, draw a spiral in the upper part of the contour.

10 Draw parallel curved lines above the clockwise spiral created in step 9.

11 Draw parallel curved lines below the clockwise spiral created in step 9.

12 Then, draw a spiral in the lower part of the contour.

13 Draw parallel curved lines below the counterclockwise spiral created in step 12.

14 Lastly, draw parallel curved lines above the clockwise spiral created in step 12 to complete the pattern.

PART 3

作品設計

作品設計是基礎技巧、初級構圖、設計概念三者的綜合體，
不只驗收成果，更是秀出創意的好機會！

A finished design sketch is the mutual product of basic skill,
elementary composition, and concept of design.
Not only are you able to reap what you have sown,
you can show your creativity to others!

作品設計 ——

The
Daisy Design

步驟説明

1

先擠出一滴顏料，往下畫
出第一條弧線。

2

將第一條弧線往右上拉，
尾端收成螺旋。

3

往右上畫出第二個螺旋，
延長中段線條，尾端收成
螺旋。

4

往右畫出第三個螺旋，完
成第一枝花莖。

5

在第一枝花莖頂端的 V 形
空隙，以弧線畫出花萼。

6

在花萼上方畫出第一片花
瓣。

7

依序畫出五片花瓣，須注
意大小與排列位置。

8

以花萼為起點，在花瓣上
依序拉出五條花蕊。

Step by step

1 Squeeze out a small amount of Henna paste and draw a downward curve.
2 Extend the curve in the upper right direction and end the curve with an inward spiral.
3 Draw a 2nd spiral in the upper right direction, elongate the line and end with another inward spiral.
4 Draw a 3rd spiral to the right to complete the 1st flower stem.
5 In the v-shape at the top of the 1st stem, draw 2 curves to form the calyx.
6 Draw the 1st petal onto the calyx as shown in the diagram.
7 And draw the rest 4 petals accordingly; take note of the size and alignment of the petals.
8 Starting from the calyx, draw a line in the center of each petal to form the pistil.

9

在第二、三片花瓣中間，
拉出一條弧線。

10

將步驟 9 的弧線尾端收成
螺旋，再往右畫一個螺旋，
完成第二枝花莖。

11

在第二枝花莖頂端的 V 形
空隙，以弧線畫出花萼，
再於花萼上方畫出花瓣。

12

依序畫出五片花瓣，須注
意大小與排列位置。

13

以花萼為起點，在花瓣上
依序拉出五條花蕊。

14

在第一枝花莖尾端的螺旋
交接處，點出圓形花心。

15

距離花心 0.5 公分處，由
下往上拉出花蕊，須與花
心稍微隔開。

16

依序拉出五條花蕊，須注
意長短與排列位置。

17

在步驟 16 的花蕊左側點出
小圓點，裝飾後段花莖。

18

最後將小圓點依序點至花
莖尾端即可。

9 Extend a curve from the middle of the 2nd and 3rd petal as shown in the diagram.
10 End the curve in step 9 in a spiral, and then draw another spiral to the right to create the 2nd stem.
11 In the V-shape at the top of the 2nd stem, draw a curve to form the calyx and draw petals on top.
12 Drawn 5 petals and take note of the size and alignment of the petals.
13 Starting from the calyx, draw a line in the center of each petal to form the pistil.
14 Under the junction between the bottom of 1st stem and the spiral, draw a dot representing the core of a flower.
15 Draw the pistils in the bottom-up direction while maintaining a 0.5 cm space with the core as shown in the diagram.
16 Draw 5 pistils, taking note of the size and alignment
17 Draw dots to the right side of the pistils drawn in step 16 to add detail to the lower part of the stem.
18 Continue the dots all the way to the end of the stem to complete design.

作品設計 ——

The Peacock Design

步驟説明

1

先擠出顏料畫變形蟲輪廓，
線條尾端與頂點密合，呈
現孔雀的頭與身體。

2

以孔雀頭部為起點，往前
畫出第一條弧線。

3

以孔雀頭部為起點，往前
畫出第二條弧線，上下密
合成鳥嘴，尾端則做出水
滴效果。

4

在孔雀頭上依序畫出三滴
加長水滴，作為頭羽。

5

以水滴畫出孔雀眼睛。

6

以孔雀頸部為起點，依序
往下畫出直線。

7

以步驟6的直線為起點，
往下畫出橢圓圖案。

8

沿著橢圓圖案內側，再畫
出水滴形狀，以顏料補滿。

Step by step

1 First squeeze out the Henna paste and draw an amoeba outline, connect the ends to form the peacock's head and body contour.
2 Starting from the peacock's head, draw a curved line to the right.
3 Starting from the peacock's head, draw a second curved line. Connect the ends to form the beak, and create a drop shape at the tip of the beak.
4 Draw 3 elongated drop shapes on the top of the head contour to create the feathers.
5 Draw a drop shape inside the head contour to form the eye.
6 Starting from the neck, draw stripes to fill in the neck area as shown in the diagram.
7 Starting from the last stripe in step 6, draw a downward oval shape.
8 Draw a drop shaped contour inside the oval, and fill it in with Henna past.

9

以孔雀身體中段為起點，
畫出兩條平行的弧線。

10

以下方弧線為起點，往上
畫出曲線。

11

以曲線環繞孔雀的身體，
須注意線條與間隔。

12

以孔雀身體尾端為起點，
畫出左旋。

13

沿著螺旋外側，往右畫出
圓圈花紋。

14

在圓圈花紋之間，依序畫
出小圓點。

15

以孔雀身體尾端為起點，
以右旋往外畫出尾羽。

16

最後依序往上畫出尾羽，
畫至所需長度後，即可完
成作品。

9 Draw 2 parallel lines at the mid-section of the body contour.
10 Starting from the lower line, draw vertical curves from the bottom-up.
11 Draw numerous vertical curves to fill the space between the parallel lines, and mind the lines and gap while doing so.
12 Starting from the bottom of the peacock's body, draw a clockwise spiral.
13 Draw rounded shapes along the outline of the spiral, going from left to right.
14 Draw dots on each of the joining points of the rounded shapes as shown in the diagram.
15 Starting from the bottom of the peacock's body, draw an outward counterclockwise spiral to create the tail feather.
16 Lastly, draw from the bottom-up as many tail feathers as needed to finish the design.

作品設計 ——

火鳥之舞
The Dancing of the Fire-bird Design

步驟説明

1

先擠出一滴顏料，往左畫出半圓弧線。

2

將步驟 1 的弧線尾端往右延伸，畫出半圓弧線。

3

沿著步驟 2 的線條弧度，畫出第二條曲線。

4

以第二條曲線頂端為起點，往上畫出第三條圓弧線。

5

以第三條圓弧線的頂端為起點，往右畫出圓弧線，尾端收成勾形，作品主體完成。

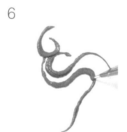

6

加粗作品主體的線條，顏料須分布均勻。

7

以第三條圓弧線下方為起點，沿著第二條曲線的弧度，往下畫出兩條弧線。

8

加粗弧線線條，顏料須分布均勻。

Step by step

1　First squeeze out the Henna paste and draw a u-shaped curve.
2　Elongate the tail of the curve from step 1 to the right to form a semi-circle.
3　Draw another curve along the first curve drawn in step 2.
4　Starting from the tip of the second curve, draw a third u-shaped curve.
5　Hook the end of the third curve upward to form a curve on the right. The 3 lines create the curved and hooked main construction.
6　Thicken the main constructions' lines, making sure the Henna paste is evenly distributed.
7　Starting from the base of the third curve, draw 2 downward curved lines parallel to the second curve.
8　Thicken the curved lines, making sure the Henna paste is evenly distributed.

9

以第二條曲線的轉折處為起點，往右畫出帶勾形的弧線。

10

以第二條曲線轉折處為起點，沿著步驟 9 的線條弧度，往右畫出波浪形曲線。

11

以第二條曲線轉折處為起點，在步驟 10 的線條下方，往下畫出波浪形曲線。

12

以第二條曲線的轉折處為起點，在步驟 11 的線條下方，往下畫出較長的波浪形曲線。

13

加粗步驟 9-12 的線條，顏料須分布均勻。

14

以第三條弧線的中段為起點，往左畫出略帶勾形的弧線。

15

沿著步驟 14 的線條弧度，依序往下畫出兩條弧線。

16

最後以顏料加粗步驟 14 的線條，即可完成作品。

9　Starting from the turning point of the second curve, draw a hooked curve to the right as shown in the diagram.

10　Starting from the turning point of the second curve, draw a waved line to the right parallel to the hooked curve in step 9.

11　Starting from the turning point of the second curve, draw another downward waved line under the waved line drawn in step 10.

12　Starting from the turning point of the second curve, draw an elongated downward waved curve under the waved line drawn in step 11.

13　Thicken the lines drawn in steps 9 through 12, making sure the Henna paste is evenly distributed.

14　Starting from the mid-point of the third curve, drawn a slightly hooked curve to the left.

15　Draw 2 more curves under and along the curve drawn in step 14.

16　Lastly, thicken the line drawn in step 14 to finish the design.

作品設計 ——

雙魚 The Dual Fish Pattern

步驟說明

1

先擠出一滴顏料，在構圖左側往下畫出尾端帶勾形的波浪曲線。

2

以波浪曲線的頂端為起點，往下畫出弧線。

3

弧線畫至尾端時往內收，再往外拉出弧度，與波浪曲線的尾端密合，完成第一條魚的輪廓。

4

在輪廓頂端往內畫出圓圈，完成魚眼。

5

在魚眼下方畫出半圓弧線。

6

在步驟 5 的弧線下方，畫出平行略長的半圓弧線。

7

在步驟 5、6 的弧線之間依序畫出直線，須注意排列的弧度與間隔。

8

沿著步驟 6 的弧度，再畫出半圓弧線。

Step by step

1　Squeeze out a drop of Henna paste, and draw a vertical waved line with a hook at the end in the left portion of the construction.

2　Starting from the tip of the previous wave, draw a downward curve as shown.

3　Draw the curve tail slightly inward, and then pull outward to form a smaller curve at the bottom, and meet the end of the line to form the tail of the first fish contour as show in the diagram.

4　Draw an inner circle at the top of the fish contour to create the fish eye.

5　Draw a rounded curve under the fish eye.

6　Draw another rounded curve under and parallel to the one drawn in step 5.

7　Draw stripes between the 2 curves drawn in steps 5 and 6, take note of the alignment with the curves and the spaces in between the stripes.

8　Draw another rounded curve under the one drawn in step 6.

9   10   11   12

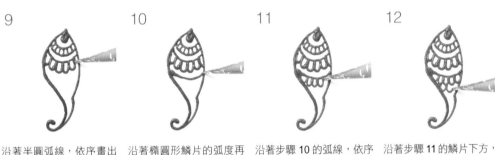

沿著半圓弧線，依序畫出橢圓形鱗片，須注意大小與排列弧度。

沿著橢圓形鱗片的弧度再畫出弧線，須注意與鱗片的間隔。

沿著步驟 10 的弧線，依序畫出橢圓形鱗片，須注意大小與排列弧度。

沿著步驟 11 的鱗片下方，畫出第二排橢圓形鱗片。

13   14   15   16

沿著第二排橢圓形鱗片下方，畫出第三排橢圓形鱗片。

沿著第三排橢圓形鱗片下方，畫出剩下的鱗片，須填滿魚身。

沿著左側輪廓，依序由下往上畫出加長水滴，完成左側鬍鬚。

沿著右側輪廓，依序由下往上畫出加長水滴，完成右側鬍鬚。

9   Draw oval-shaped fish scales under and along the rounded curve drawn in step 8, take note of the scale size and alignment.

10   Under the oval-shaped scales, draw another rounded curve. Remember to maintain a slight distance with the scales as shown in the diagram.

11   Draw oval-shaped fish scales under and along the rounded curve drawn in step 10, take note of the scale size and alignment.

12   Draw a second row of oval-shaped fish scales under and along the scales drawn in step 11.

13   Draw a third row of oval-shaped fish scales under and along the scales drawn in step 12.

14   Draw another row of oval-shaped fish scales under and along the scales drawn in step 13, keep drawing scales to fill the fish contour as needed.

15   Along the left profile of the fish contour, draw 2 elongated drop shapes from the bottom up to finish the left whiskers.

16   Along the right profile of the fish contour, draw 2 elongated drop shapes from the bottom-up to finish the right whiskers.

17	18	19	20

在魚身尾端由下往上畫出水滴。	依序畫出水滴完成魚尾，完成第一條魚，須注意水滴長短與排列弧度。	在第一條魚的右下方，往下畫出尾端帶勾形的波浪曲線。	以波浪曲線的頂端為起點，往下畫出弧線，與波浪曲線的尾端密合。

21	22	23	24

在輪廓頂端往內畫出圓圈與橢圓點，完成魚眼。	在魚眼下方畫出半圓弧線。	在步驟 22 的弧線下方，畫出平行的半圓弧線。	沿著步驟 23 的弧線，依序畫出橢圓形鱗片，須注意大小與排列弧度。

17　Draw elongated drop shapes from the bottom-up at the end of the fish contour.

18　Draw as many elongated drop shapes as needed to complete the fish tail and to finish the first fish. Remember to mind the length and alignment of the drop shapes.

19　At the lower right corner of the construction, draw a vertical waved line with a hook at the end.

20　Starting from the tip of the previous wave, draw a downward curve, and meet the end of the line to form the tail of the second fish contour as show in the diagram.

21　Draw an inner circle at the top of the fish contour to create the fish eye.

22　Draw a rounded curve under the fish eye.

23　Draw another rounded curve under and parallel to the one drawn in step 22.

24　Draw oval-shaped fish scales under and along the rounded curve drawn in step 23, take note of the scale size and alignment.

25

沿著步驟 24 的鱗片下方，
往下依序畫出弧線填滿魚
身。

26

沿著右側輪廓，依序由下
往上畫出加長水滴，完成
右側鬍鬚。

27

在魚身尾端由下往上畫出
水滴。

28

依序畫出水滴完成魚尾，
完成第二條魚，須注意水
滴長短與排列弧度。

29

以魚嘴上方為起點，擠出
小圓點。

30

沿著左側輪廓，繼續擠出
小圓點。

31

將小圓點依序擠至魚身尾
端，再往左下方擠出，完
成氣泡圖案後，即可完成
作品。

25　Under the oval-shaped scales, draw another rounded curve, and keep drawing until the fish contour is filled.

26　Along the right profile of the fish contour, draw 2 elongated drop shapes from the bottom-up to finish the right whiskers.

27　Draw elongated drop shapes from the bottom-up at the end of the fish contour.

28　Draw as many elongated drop shapes as needed to complete the fish tail and to finish the second fish. Remember to mind the length and alignment of the drop shapes.

29　Starting at the top of the second fish contour, draw small dots as shown in the diagram.

30　Keep on drawing small dots along the left side of the second fish.

31　Extend the dots to the tail of the second fish, and curve the alignment of the dots in the lower left direction to complete the bubbles and finish the design.

作品設計 ——

The Insignia Design ——

步驟説明

1

先擠出一滴顏料，在構圖左側畫出弧線。

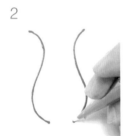

2

在構圖右側畫出與步驟 1 方向相反的弧線，須預留空間。

3

在兩條弧線之間擠出圓點，再以圓點為起點，往左上畫出弧線。

4

將步驟 3 的弧線尾端收成螺旋，右側重複步驟 3-4。

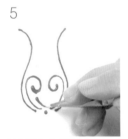

5

以步驟 3 的螺旋下方為起點，往右下畫出第一滴加長水滴，須注意與螺旋和底下圓點的距離。

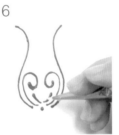

6

以步驟 4 的螺旋下方為起點，往左下畫出第二滴加長水滴，須注意與螺旋和底下圓點的距離。

7

第三滴加長水滴重複步驟 5。

8

第四滴加長水滴重複步驟 6。

Step by step

1 Squeeze out a drop of Henna paste, and draw a vertical curve in the left portion of the construction.
2 In the right portion, draw a vertical curve symmetrical to the one drawn in step 1. Remember to maintain a small distance between the curves as shown in the diagram.
3 Draw a dot between the ends of the curves. Use the dot as a starting point and draw a curve in the upper left direction.
4 Extend the curve in step 3 into a spiral, and repeat steps 3 and 4 on the opposite side.
5 Starting from under the spiral drawn in step 3, draw an elongated drop shape in the lower right direction.
6 Starting from under the spiral drawn in step 4, draw a second elongated drop shape in the lower left direction. Take note of the drop shapes' distances between the spirals and the dot.
7 Repeat step 5 to create the third elongated drop shape.
8 Repeat step 6 to create the fourth elongated drop shape.

在第三、四滴加長水滴之間，畫出第五滴加長水滴，須注意水滴間的距離與長度。

以第五滴加長水滴的上方為起點，依序擠出三個圓點裝飾，完成作品底座。

沿著步驟 1 的弧度，在內側往下畫出弧線。

沿著步驟 2 的弧度，在內側往下畫出弧線。

在步驟 11 的弧線內側，擠出圓點裝飾。

以圓點依序裝飾弧線內側，須注意大小與間隔。

在步驟 12 的弧線內側，擠出圓點裝飾。

以圓點依序裝飾弧線內側，須注意大小與間隔。

9 　Draw a fifth elongated drop shape between the third and fourth one, take note of the length and distances of the drops.

10 　Starting from the fifth elongated drop shape, draw 3 dots in a row to decorate, thus completes the base portion of this design.

11 　Draw a downward curve along the inner profile of the curve drawn in step 1.

12 　Draw a downward curve along the inner profile of the curve drawn in step 2.

13 　Decorate the curve drawn in step 11 with a row of dots.

14 　Finish the row of dots, and mind the size and spaces in between.

15 　Decorate the curve drawn in step 12 with a row of dots.

16 　Finish the row of dots, and mind the size and spaces in between.

17

沿著步驟 **1** 的弧度，在外側往上畫出左旋藤蔓。

18

沿著步驟 **2** 的弧度，在外側往上畫出右旋藤蔓。

19

在步驟 **14** 的線條上方，往上畫出小的左旋藤蔓。

20

在步驟 **16** 的線條上方，往上畫出小的右旋藤蔓。

21

在小的左右藤蔓之間，擠出圓點裝飾。

22

以步驟 **21** 的圓點上方為起點，畫出加長水滴。

23

在加長水滴的左側，往右下方畫出加長水滴。

24

在步驟 **23** 的水滴左側，再依序往右下方畫出兩滴加長水滴，須注意長度與間隔。

17 Draw an upward clockwise spiral along the outer profile of the curve drawn in step 1.
18 Draw an upward counterclockwise spiral along the outer profile of the curve drawn in step 2.
19 On the line created in step 14, draw a small upward clockwise spiraled vine.
20 On the line created in step 16, draw a small upward counterclockwise spiraled vine.
21 Draw a decorative dot between the spiraled vines.
22 Starting from the dot drawn in step 21, draw an upward elongated drop shape.
23 Then, draw a drop shape in the lower right direction to the left of the drop shape drawn in step 22.
24 Draw 2 more drop shapes to the left of the drop shape drawn in step 23 in a fan like matter, take note of the length and spacing of the drop shapes.

25 	26 	27 	28 

在步驟 22 的加長水滴右側，往左下方畫出加長水滴。

在步驟 25 的水滴右側，再依序往左下方畫出兩滴加長水滴，須注意長度與間隔。

在作品底部擠出顏料，往右拉出較大的加長水滴。

在作品底部擠出顏料，往左拉出較大的加長水滴。

29 	30 	31 	32 

沿著步驟 27 的弧度，再拉出較大的加長水滴。

沿著步驟 28 的弧度，再拉出較大的加長水滴。

依序拉出底部的水滴，須注意長度、間隔，與外圍弧度。

最後往上拉出中間的直線水滴，即可完成作品。

25 Draw a drop shape in the lower left direction to the right of the drop shape drawn in step 22.

26 Draw 2 more drop shapes to the right of the drop shape drawn in step 25 in a fan like matter, take note of the length and spacing of the drop shapes.

27 Squeeze some Henna paste at the base of the design and extend a slightly bigger drop shape to the right as shown in the diagram.

28 Squeeze some Henna paste at the base of the design and extend a slightly bigger drop shape to the left as shown in the diagram.

29 Draw another slightly bigger elongated drop shape below and along the drop shape drawn in step 27.

30 Draw another slightly bigger elongated drop shape below and along the drop shape drawn in step 28.

31 Finish drawing the rest of the elongated drop shapes at the base accordingly; take note of the length, spacing, and curviness of the drop shapes.

32 Lastly, draw the vertical middle drop shape from the bottom-up to finish the design.

作品設計 ——

The Floral Design with Vines

步驟説明

1

先擠出一滴顏料，畫出左旋藤蔓。

2

以左旋藤蔓尾端為起點，往上拉出加長水滴。

3

沿著步驟 2 的弧度，往外側拉出較短的加長水滴。

4

以左旋藤蔓尾端為起點，往右畫出右旋，完成花梗。

5

在左旋藤蔓與右旋之間，畫出兩條弧線，完成花萼。

6

在花萼上方，依序畫出圓形花蕊，須注意大小。

7

以花萼左側為起點，往上畫出心形花瓣。

8

依序畫出心形花瓣，須注意大小與排列弧度。

Step by step

1 Squeeze out a drop of Henna paste, and draw a clockwise spiraled vine.
2 Starting from the end of the clockwise spiraled vine, extend the stroke upward to form an elongated drop shape.
3 Extend another shorter elongated drop shape to the left and follow the curviness of the elongated drop drawn in step 2.
4 Starting from the bottom of the counterclockwise spiraled vine, draw a clockwise spiral to the right to form the flower base.
5 Draw 2 rounded curves between the clockwise spiraled vine and the counterclockwise spiral to form the calyx.
6 Starting from the calyx, draw rounded shapes along the curve to form the pistil, take note of the size.
7 Starting from the left side of the calyx, draw an upward heart-shaped petal.
8 Draw the remaining 2 petals accordingly; take note of the size and alignment of the petals.

9

依序畫出心形花瓣，須注
意大小與排列弧度。

10

依序畫出拉線花蕊。

11

在左旋與右側花瓣之間，
往下畫出尾端帶螺旋的弧
線。

12

以左旋下方為起點，往下
畫出弧線，並與步驟 11 的
弧線密合，完成葉子輪廓。

13

沿著步驟 11 的弧度，在
葉子輪廓內畫出弧線。

14

在葉子左側，以弧線依序
畫出葉脈。

15

在葉子右側，以弧線依序
畫出葉脈。

16

在右側花瓣與葉子之間，
依序畫出兩條水滴形線條。

17

在右側花瓣與葉子之間，
往上畫出左旋。

18

在右旋與水滴線條之間，
畫出右旋藤蔓。

9 Use gentle strokes to draw stamens within the petals, take note of the length and spacing.
10 Finish drawing the stamens accordingly as shown in the diagram.
11 Starting from between the clockwise spiral and the right-side petal, draw a downward curve with a spiraled end.
12 Starting from the lower part of the clockwise spiral, draw a downward curve, and joint ends with the curve drawn in step 11 to finish the outline of the leaf.
13 Follow the curviness of the curve drawn in step 11 to draw the central leaf vein in the leaf contour.
14 Draw the branched out veins on the left side all the way down the central vein.
15 Draw the branched out veins on the right side all the way down the central vein.
16 Starting from between the right-side petal and the leaf, draw 2 elongated drop shapes.
17 Starting from between the right-side petal and the leaf, draw an upward clockwise spiral.
18 Starting from between the counterclockwise spiral and the elongated drop, draw a counterclockwise spiraled vine.

19 	20 	21 	22 
以右旋頂端為起點，沿著右旋藤蔓的弧度，往上畫出尾端帶勾形的弧線。	沿著步驟 19 的弧度，在線條左側往上畫出兩條不等長的水滴形線條，完成作品右側藤蔓，須注意長度與距離。	在左側花瓣與左旋藤蔓之間，往上畫出右旋線條。	在左旋藤蔓與右旋線條之間，往上畫出尾端帶勾形的弧線。
23 	24 	25 	26 
以步驟 22 的線條尾端為起點，往上畫出半圓形圖案，須注意大小與間隔。	在半圓形圖案中間，依序點出圓點裝飾。	在步驟 21、22 的線條頂端之間，畫出四滴水滴裝飾，須注意大小與排列弧度。	最後以步驟 22 的線條上方為起點，依序擠出圓點裝飾即可。

19　Starting from the top of the counterclockwise spiral, follow the curviness of the vine and draw an upward curve with a hook on the end.

20　Follow the curviness of the curve drawn in step 19 and draw 2 elongated drop shapes with uneven lengths to the left, thus finishes the design's right-sided vines. Remember to mind the length and distance of the curves.

21　Starting from between the left-side petal and the counterclockwise spiral, draw an upward curve with a counterclockwise spiral.

22　Starting from between the clockwise spiraled vine and the counterclockwise spiral, draw an elongated upward curve with a hook at the end.

23　Starting from the same point as step 22, draw semi-circled patterns up and along the curve drawn in step 22, mind the sizes and spacing.

24　Dot all the semi-circles for decoration.

25　Between the tops of the curves drawn in step 21 and 22, draw 4 elongated drop shapes for decoration; take note of the sizes and curviness alignment.

26　Lastly, decorate the top of the curve drawn in step 22 with dots to complete the design.

作品設計 ——

The Assembled Leaves Design

步驟說明

1	2	3	4

1 先擠出一滴顏料，畫出波浪曲線。

2 將波浪曲線尾端收成螺旋。

3 以波浪曲線中段為起點，往右下畫出帶勾形的波浪曲線。

4 以步驟2的螺旋為起點，往下畫出弧線，與勾形尾端密合。

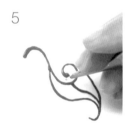   

5 以勾形尾端為起點，往左上畫出曲線，完成第一片葉子的輪廓。

6 在葉子左側畫出直線葉脈。

7 以葉脈依序填滿葉子左側，須注意間隔與排列弧度。

8 在葉子右側畫出直線葉脈，依序填滿葉子右側，須注意間隔與排列弧度。

Step by step

1 Squeeze a drop of Henna paste and draw a waved curve.
2 Extend the tail of the curve into a spiral.
3 Starting from the mid-point of the waved curve, draw a downward waved curve in the lower right direction, and end the curve with a hook.
4 Starting from the spiral drawn in step 2, draw a downward curve and join the end with the hook drawn in the previous step as show in the diagram.
5 Using the hooked tip as a starting point, draw a curve in the upper left direction, thus completes the contour of the first leaf.
6 Draw a straight vein line on the left of the leaf.
7 Finish drawing all the left leaf veins, take note of the spacing and curviness alignment.
8 Then, draw the right leaf veins; take note of the spacing and curviness alignment.

9 	10 	11 	12 
以葉柄中段為起點，往左畫出右旋，往右畫出圓弧線。	以葉柄中段為起點，往上畫出波浪形曲線，再往右上拉出弧線。	將步驟 10 的弧線頂端與波浪形曲線的頂端密合，完成第二片葉子的輪廓。	以第二片葉子的輪廓尾端為起點，往上畫出直線。

13 	14 	15 	16 
依序以直線、弧線畫出左右側葉脈，填滿葉子輪廓，須注意間隔與排列弧度。	在第二片葉子右側，依序往上畫出帶勾形弧線與加長水滴，完成第二片葉子構圖。	以葉柄中段為起點，沿著第一片葉子的左側弧度，往下畫出加長水滴。	沿著步驟 15 的弧度，在左側往下畫出較短的加長水滴。

9 Starting from the mid-point of the leaf stem, draw a counterclockwise spiral to the left, and a clockwise spiral to the right.

10 Starting from the same point as step 9, draw an upward waved curve, and another upward curve in the upper right direction.

11 Join the ends of the curves in step 10 to form the contour of the second leaf.

12 Draw a straight line starting from the base of the second leaf's contour up to the tip of the leaf.

13 Use straight and curved lines to draw the right leaf veins, take note of the spacing and curviness alignment.

14 Draw a hooked curve and an elongated drop shape to the right of the second leaf, thus finishes the construction of the second leaf.

15 Starting from the mid-point of the leaf stem, draw an elongated drop shape along the left profile of the first leaf as shown in the diagram.

16 Follow the curve of the elongated drop shape drawn in step 15, and draw another shorter elongated drop shape downward to its left.

17	18	19	20

以第一片葉子的螺旋為起點，往右下畫出螺旋在下的藤蔓。	以第一片葉子的螺旋為起點，沿著步驟 17 的弧度，往右下畫出弧線，以水滴技法收尾。	以步驟 18 的弧線中段為起點，往右重複步驟 3-4 後，往上畫出加長水滴，再於輪廓內側畫出直線，完成第三片葉子的基本構圖。	依序以直線畫出左右側葉脈，填滿葉子輪廓，須注意間隔與排列弧度。

21	22	23	24
		23	24
以第三片葉子下方為起點，往右下畫出弧線，以圓點收尾。	沿著步驟 21 的弧度，在上方往右下畫出弧線，以圓點往上收尾。	以第三片葉子下方為起點，沿著葉子弧度，往上畫出弧線，以圓點收尾。	最後以步驟 22、23 之間為起點，往右上畫出帶勾形的弧線，再以圓點收尾即可。

17 Starting from the first leaf's spiral, draw a downward spiraled vine in the lower right direction.

18 Starting from the first leaf's spiral, follow the curve of the vine drawn in step 17 to draw a curve in the lower right direction, ending with a drop shape technique as shown.

19 Starting from the mid-point of the curve drawn in step 18, repeat steps 3 through 4 to the right, and extend the lines upward into an upward curled tail. Then, draw a line down the middle of this contour, thus completes the construction of the third leaf.

20 Then, draw the leaf veins on both sides to fill the entire leaf contour; take note of the spacing and curviness alignment.

21 Using the bottom of the third leaf as the starting point, draw a downward curve in the lower right direction. End the curve with a rounded end as shown in the diagram.

22 Along the curve drawn in step 21, draw another downward curve in the lower right direction. End the curve with a rounded upward end.

23 Using the bottom of the third leaf as the starting point, draw an upward curve following the curve of the third leaf; finish the curve with a rounded end.

24 Lastly, use the junction between the curves created in steps 22 and 23 as a starting point and draw a hook-tailed curve in the upper right direction, finish the curve with a rounded end to complete the design.

作品設計 ————

珠寶飾紋
The Jewel Decoration Pattern

步驟説明

1	2	3

先擠出一滴顏料，在構圖
中心畫出一個圓圈。

在圓圈外圍再畫一個大圓，
成一同心圓圖案。

沿著同心圓弧度，依序畫
出圓形花瓣。

4	5	6

將花瓣繞滿同心圓，須注
意間隔與大小。

以花瓣頂端為起點，往上
畫出橢圓形圓點。

重複步驟 5，須注意間隔
與大小。

Step by step

1　Squeeze a drop of Henna paste and draw a circle in the middle of the construction.
2　Draw another bigger circle around the original circle to form concentric circles.
3　Draw rounded petals along the boarder of the circle.
4　Finish drawing petals until completing the full circle; mind the spacing and size of the petals.
5　Starting from the top of the petals, draw upward oval dots as shown in the diagram.
6　Repeat step 5, and mind the spacing and size of the dots.

以橢圓形圓點為起點，往上畫出水滴圖案。

重複步驟 7，須注意間隔與大小。

以不帶橢圓形圓點的花瓣頂端為起點，往上畫出水滴圖案。

以步驟 9 的水滴圖案為起點，往上畫出直線裝飾。

重複步驟 9-10，依序畫出圖案。

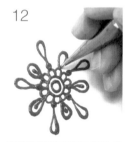

沿著花瓣弧度畫完步驟 10 的圖案後，即可完成。

7　Using the oval dots as the starting point, draw water drop contours.
8　Repeat step 7, and mind the spacing and size of the water drop contours.
9　On the oval dots that do not have oval dots above them, draw drop shaped contours.
10　Draw straight lines within the drop shaped contours for decoration as shown in the diagram.
11　Repeat steps 9 through 10 and draw the pattern accordingly.
12　Finish the full circle with patterns according to step 10 to complete the entire pattern.

作品設計 ——

緞帶飾紋
The Ribbon Pattern

步驟説明

1

先擠出一滴顏料,往右畫出直線。

2

以直線左側為起點,往右畫出橢圓形花紋。

3

將橢圓形花紋畫至所需長度,須注意大小與間隔。

4

以橢圓形花紋左側為起點,往上畫出半圓弧線。

5

將半圓弧線尾端往上收成螺旋。

6

以橢圓形花紋上方為起點,在螺旋右側往下畫出右旋,須注意兩者的距離。

7

以橢圓形花紋上方為起點,在右旋右側往下畫出左旋,須注意兩者的距離。

8

將左旋尾端往上收起,再往下畫出半圓弧線。

Step by step

1 Squeeze a drop of Henna paste and draw a straight line to the right.

2 Starting from the left end of the line, draw semi-oval shapes along the line to the right.

3 Draw the semi-ovals as needed, take note of the size and spacing of the semi-ovals.

4 Starting from the left end of the semi-ovals, draw an upward semi-circle.

5 Extend the tail of the semi-circle into a spiral.

6 Starting from above the row of semi-ovals, draw a downward counterclockwise spiral to the right of the spiral drawn in step 5, take note of the distance between spirals.

7 Starting from above the row of semi-ovals, draw a downward clockwise spiral to the right of the counterclockwise spiral drawn in step 6, take note of the distance between spirals.

8 Then extend the tail of the clockwise spiral upward, and then turn downward to form a semi-circle curve as shown in the diagram.

9	10	11	12

以橢圓形花紋上方為起點，在半圓弧線右側往下畫出右旋。

以橢圓形花紋上方為起點，在步驟 9 的右旋右側往下畫出左旋，再將尾端往上收起。

將步驟 10 的尾端往下以左旋線條收尾，並以顏料填滿。

在步驟 8、5 的半圓形內側，畫出略小的半圓輪廓，並以顏料填滿。

13	14	15	16

以橢圓形上方花紋為起點，在步驟 5、6 之間擠出圓點。

以步驟 13 的圓點為起點，往上畫出水滴圖案。

以顏料填滿步驟 14 的水滴圖案。

在步驟 6 的右旋頂端，擠出圓點裝飾。

9 Starting from above the row of semi-ovals, draw a downward counterclockwise spiral to the right of the semi-circle curve drawn in step 8.

10 Starting from above the row of semi-ovals, draw a downward clockwise spiral to the right of the counterclockwise spiral drawn in step 9, and extend the tail upward.

11 Retract the tail drawn in step 10 downward into a clockwise spiral, and use Henna paste to fill it in as shown in the diagram.

12 In the semi-circles drawn in steps 8 and 5, draw a slightly smaller semi-circle, and then fill them in with Henna paste.

13 Starting from above the row of semi-ovals, draw a dot between the curve and the spiral created in steps 5 and 6 as shown in the diagram.

14 Starting from the dot drawn in step 13, draw an upward water drop shape.

15 Use the Henna paste to fill in the water drop shape drawn in step 14.

16 Draw a dot on top of the counterclockwise spiral drawn in step 6 for decoration.

17	18	19	20

在步驟 8、9 的圖案之間，重複步驟 13-15。

在步驟 9 的右旋頂端，擠出圓點裝飾，完成作品上半部。

在作品下半部往右畫出一條平行等長的直線。

以步驟 19 的左側為起點，依序往右畫出橢圓形花紋，須注意大小與間隔。

21	22

以橢圓形花紋左側為起點，兩兩之間往下依序擠出橢圓圓點裝飾，須注意大小。

最後將橢圓圓點擠至所需長度，即可完成作品。

17 Repeat steps 13 trough 15 between the patterns drawn in steps 8 and 9.

18 Draw a dot on top of the counterclockwise spiral drawn in step 9 for decoration.

19 Draw a parallel line under the pattern as shown.

20 Starting from the left end of the line drawn in step 19, draw upside-down semi-oval shapes along the line to the right, take note of the size and spacing of the semi-ovals.

21 Starting from the left end of the upside-down semi-ovals, draw oval dots on top of every other upside-down semi-oval, and take note of the oval dot size.

22 Finish drawing the oval dots as needed to complete the pattern.

作品設計 ——

The Sun Design

步驟說明

1

先擠出一滴顏料，在構圖中心畫出一個右旋。

2

右旋線條畫至尾端時，以波浪曲線往左延伸。

3

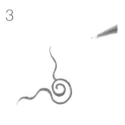

以右旋上方為起點，往上畫出第二條波浪曲線。

4

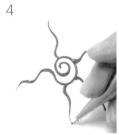

沿著右旋的弧度，依序畫出第三至五條波浪曲線，讓圖案呈放射形，須注意間隔與長度。

5

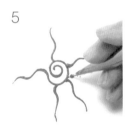

在第三、四條曲線之間，擠出圓點圖案。

6

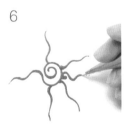

將圓點往右畫出小圓弧線，尾端收成波浪曲線。

7

重複步驟6，依序將曲線間隔填滿後，即可完成圖案。

Step by step

1 First squeeze out the Henna paste and draw a counterclockwise spiral in the middle of the construction.

2 At the end of the spiral, extend the tail to the left in a waved line.

3 Starting from the top of the counterclockwise spiral, draw a 2nd waved line in an upward direction.

4 Draw the 3rd to 5th waved lines around the spiral as shown in the diagram in a radial matter, take note of the length and spaces between the waved lines.

5 Squeeze a small drop of Henna paste between the 3rd and 4th line.

6 Starting from the drop, draw a small rounded curved line to the right and finish with a waved line.

7 Repeat step 6 and finish all the gaps accordingly to complete the design.

作品設計 ──

原住民
The Aboriginal Pattern
飾紋

步驟說明

1　先擠出顏料，畫出第一組上下平行的直線。

2　在平行線間擠出一滴圓點顏料，再於右側畫出右旋，須注意間隔距離。

3　重複步驟 2，畫至所需的長度。

4　在步驟 3 的圖案下方，畫出第二組上下平行的直線。

5　在平行線間畫出 X 形圖案。

6　畫 X 形圖案時須注意筆順，先將第一條線往右下收。

7　將 X 形圖案的第二條線往右上拉。重複步驟 5-7，畫至需要的長度後，第二組平行線間呈菱形與三角形的組合。

8　在倒三角形內側畫出半圓形輪廓。

Step by step

1　First squeeze out the Henna paste and draw a pair of horizontal parallel lines.
2　Squeeze a drop in between the lines to make a dot, and draw a counterclockwise spiral to the right of the dot, take note of the spacing.
3　Repeat step 2 until finishing the needed length.
4　Under the pattern drawn in step 3, draw a second pair of parallel lines.
5　Draw Xs in between the second pair of parallel lines.
6　Remember to mind the strokes when drawing the Xs, the first stroke should go in the lower right direction.
7　Then, the second stroke should go in the upper right direction. Repeat steps 5 through 7 as needed, the pattern in between the second pair of parallel lines should be consisted with diamonds and triangles.
8　Draw a semi-circle in the reversed triangle.

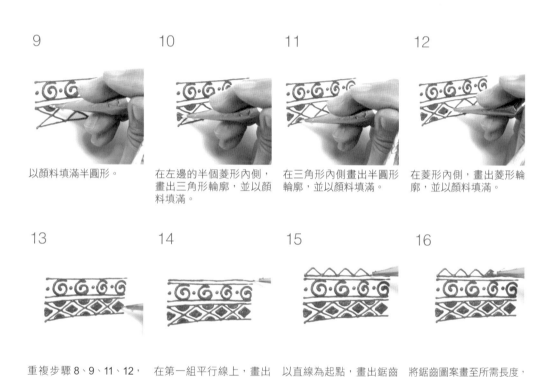

9

以顏料填滿半圓形。

10

在左邊的半個菱形內側，畫出三角形輪廓，並以顏料填滿。

11

在三角形內側畫出半圓形輪廓，並以顏料填滿。

12

在菱形內側，畫出菱形輪廓，並以顏料填滿。

13

重複步驟 8、9、11、12，畫至所需長度。

14

在第一組平行線上，畫出一條直線。

15

以直線為起點，畫出鋸齒圖案。

16

將鋸齒圖案畫至所需長度，再以顏料填滿後，即可完成作品。

9　Use the Henna paste to fill in the semi-circle.
10　Within the half of diamond shape on the left end, draw a triangle and fill it in with Henna paste.
11　Draw a semi-circle in the triangle and fill it in with Henna paste as shown in the diagram.
12　Draw a slightly smaller diamond shape in the diamond and fill it in with Henna paste.
13　Repeat steps 8, 9, 11, and 12 as needed.
14　Draw a horizontal line parallel and above to the first pair of parallel lines.
15　Draw a jagged pattern on the horizontal line drawn in step 14.
16　Finish the jagged pattern as needed, and then fill it in with Henna paste to complete the pattern.

PART 4

Body Henna

人體繪製是更高階的挑戰，
需要跟著身體線條走，
與手稿的感覺有明顯差距，
學會比例拿捏和構圖的最好方式，就是多加練習～

To transfer your designs onto the human skin is quite a challenge;
you must follow the body's natural curves,
which is quite different than sketching on paper.
The best way to gain control over the proportion and
composition of the design on the human skin is with a lot of practice.

手臂 Henna —
婚禮飾紋
Henna on the arm —
The Bridal Design

步驟説明

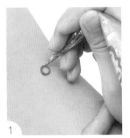

1
先擠出一滴顏料，在構圖
上方畫出圓圈。

2
沿著圓圈的弧度依序畫出
圓形花蕊，須注意大小與
間隔。

3
以圓形花蕊的間隔為起點，
沿著花蕊弧度依序畫出心
形花瓣，須注意大小與間
隔。

4
以圓形花蕊為起點，在每
片花瓣內側畫出花蕊，須
注意長度與排列位置，部
分花蕊可採刮線技法。

5
以花瓣之間為起點，往左
上畫出左旋藤蔓與帶弧形
的加長水滴。

6
沿著步驟 5 的水滴右側，往
下畫出略短的加長水滴。

7
在步驟 6 的水滴右側，往
上畫出右旋藤蔓。

8
以花瓣之間為起點，往上
畫出左旋藤蔓。

Step by step

1 First squeeze out a small drop of Henna paste, and draw a circle in the upper portion of the construction.
2 Draw rounded pistils around the circle; take note of the size and spacing of pistils.
3 Using the rounded pistils as the starting point, draw heart-shaped petals along the pistils, take note of the size and spacing of the petals.
4 Using the rounded pistils as the starting point, draw pistil within each petal, take note of the size and alignment of pistils. Partial pistils may be drawn with the flicker technique.
5 Starting from between the petals, draw a clockwise spiral and a curved elongated drop shape in the upper left direction.
6 Draw a slightly shorter downwards elongated drop shape along the elongated drop shape drawn in step 5.
7 Draw an upwards counterclockwise spiraled vine to the right of the elongated drop shape drawn in step 6.
8 Starting from between the petals, draw an upwards clockwise spiraled vine.

9
以左旋藤蔓頂端為起點，
在左側往上畫出加長水滴。

10
以左旋藤蔓底部為起點，
往右上畫出波浪曲線，尾
端收成水滴。

11
以波浪曲線的下方為起點，
往下畫出右旋藤蔓，完成
作品上方構圖。

12
以花瓣之間為起點，畫出
左旋藤蔓。

13
以花瓣之間為起點，往下
畫出弧線，尾端收成水滴。

14
沿著步驟 13 的弧度，在右
側畫出略長的弧線，尾端
收成水滴。

15
沿著步驟 14 的弧度，在右
側往下畫出弧線。

16
沿著步驟 12 的藤蔓內側，
往下畫出弧線，尾端與步
驟 15 的尾端密合，完成葉
子輪廓。

17
以葉子輪廓頂端為起點，
在內側往下畫出弧線，尾
端收尖。

9 Starting from under the tip of the clockwise spiraled vine, draw an elongated drop shape in the upper left direction.
10 Starting from the base of the clockwise spiraled vine, draw a waved curve with a drop shaped end in the upper right direction.
11 Starting from the base of the waved curve, draw a downward counterclockwise spiraled vine thus finishes the upper portion of the construction.
12 Starting from between the petals, draw a clockwise spiraled vine.
13 Starting from between the petals, draw a downward curve with a drop shaped end.
14 Following the curviness of the curve drawn in step 13, draw another slightly long curve with a drop shaped end on the right of the curve as shown in the diagram.
15 Following the curviness of the curve drawn in step 14, draw a curve in the lower right direction.
16 Draw a downward curve from the clockwise spiraled vine join the end with the curve drawn in step 15 to finish the contour of the leaf.
17 Starting from the base of the leaf, draw a sharp ended downward curve in the middle.

18
在葉柄處往下畫出加長水滴裝飾。

19
以花瓣之間為起點，往下畫出藤蔓。

20
以花瓣之間為起點，往下畫出弧線，畫至尾端時往內收，再往外拉出弧度，與藤蔓的尾端密合，完成作品下方的輪廓。

21
在輪廓內的左側，畫出一個螺旋。

22
沿著步驟 21 的螺旋弧度，往右上畫出弧線。

23
沿著步驟 22 的弧度，依序往右上畫出圓形花紋，須注意大小。

24
沿著步驟 23 的弧度，往右上畫出弧線。

25
沿著步驟 24 的弧度，往下畫出扁圓形花紋。

26
依序畫出扁圓形花紋，須注意大小。

18　Draw an elongated drop shape for decoration at the base of the leaf stem as shown.
19　Starting from between the petals, draw a downward spiraled vine.
20　Starting from between the petals, draw a downward curve and retract the tail of the curve inwards. Then, curve outwards and join the end with the clockwise spiraled vine.
21　Draw a spiral in the upper left corner of the contour just drawn.
22　Draw a curve in the upper right direction following the curviness of the spiral drawn in step 21.
23　Draw a row of small circles on the curve drawn in step 22, take note of the size.
24　Following the curviness of the circles drawn in step 23, draw a curve in the upper right direction.
25　Following the curviness of the curve drawn in step 24, draw a row of flat rounded patterns as shown.
26　Finish drawing the flat rounded patterns as needed, and take note of the size.

27
以顏料填滿輪廓尾端。

28
以輪廓左側上方為起點，往左畫出帶弧度的加長水滴。

29
沿著步驟 28 的弧度，往左畫出弧度較平的加長水滴。

30
以輪廓左側上方為起點，往下畫出較長的加長水滴。

31
以輪廓左側上方與步驟 30 的水滴之間為起點，往下畫出波浪曲線，尾端收成螺旋，長度須比輪廓長。

32
在波浪曲線的左側上，往左畫出小水滴。

33
依序畫出小水滴，須注意長度與間隔。

34
在波浪曲線的螺旋右側，往上畫出水滴裝飾。

35
依序畫出全部的水滴，須注意大小與排列的弧度。

36
以最後一滴水滴的頂端為起點，往上擠出小圓點裝飾，即可完成作品。

27 Use Henna paste to fill in the tail of the contour.
28 Starting from the upper left corner of the contour, draw a curved elongated drop shape to the left.
29 Following the curviness of the curve drawn in step 28, draw a slightly flattened elongated drop shape to the left.
30 Starting from the upper left corner of the contour, draw a downward elongated drop shape.
31 Starting from in between the upper left corner of the contour and the drop shape drawn in step 30, draw a downward spiral ended waved curve with a spiraled end, the length should surpass the contour.
32 Draw small drop shapes on the left side of the waved curve drawn in step 32.
33 Finish drawing the entire row of small drop shapes, take note of the length and spacing of the drops.
34 Draw drop shapes from the bottom-up to the right of the wave curve's spiral end.
35 Finish drawing the entire row of drop shapes, take note of the length and spacing of the drops.
36 Starting from the end of the last drop shape, draw a row of small dots upward to finish the design.

手背 Henna —
印度花飾
Henna on the hand —
The Indian Floral Design

步驟説明

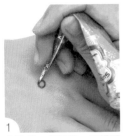

1
先擠出一滴顏料,在構圖
中心畫出圓圈。

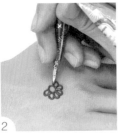

2
沿著圓圈的弧度依序畫出
圓形花蕊,須注意大小與
間隔。

3
以圓形花蕊的間隔為起點,
往上畫出心形花瓣。

4
沿著花蕊弧度依序畫出花
瓣,須注意大小與間隔。

5
以圓形花蕊頂端為起點,在
花瓣內側畫出圓點裝飾。

6
依序畫出圓點,須注意大
小與排列位置。

7
以花瓣之間為起點,往左
畫出尾端帶勾形的波浪曲
線。

8
在曲線的上方畫出弧線,
線條頂端與波浪曲線尾端
密合。

Step by step

1 First squeeze out a small drop of Henna paste, and draw a circle in the middle of the construction.
2 Draw rounded pistils around the circle, take note of the size and spacing of pistils.
3 Using the rounded pistils as the starting point, draw an upward heart-shaped petal.
4 Finish all the petals around the pistils, take note of the size and spacing of the petals.
5 Using the rounded pistils as the starting point, draw rounded dots to decorate the petal.
6 Finish drawing the dots inside each petal, take note of the size and alignment of dots.
7 Draw a hook ended waved curve to the left extending from between 2 of the petals as shown.
8 Draw a curved line on top of the waved curve drawn in step 7 and meeting the ends together.

9
以花瓣之間為起點，在波浪曲線與弧線之間往左上畫出曲線，完成第一片葉子的輪廓。

10
以弧線葉脈依序填滿葉子內側。

11
以花瓣頂端為起點，沿著葉子弧度，往左畫出弧線。

12
以步驟 11 的弧線為起點，畫出水滴。

13
沿著弧線依序畫出水滴，須注意大小與間隔。

14
以花瓣之間為起點，往下畫出螺旋。

15
以花瓣之間為起點，往下畫出帶勾形的波浪曲線。

16
以步驟 15 的曲線頂端為起點，往下畫出水滴。

17
沿著曲線依序畫出水滴，須注意大小與間隔。

18
以步驟 14 的右旋尾端為起點，往右畫出藤蔓，須延伸至手指。

9 Draw a curved line down the middle of the contour just drawn starting from the same starting point used in step 7 thus finishes the contour of the first leaf.

10 Use curved lines to fill in all the leaf vines.

11 Starting from the top of the petal, draw a curved line in the upper left direction following the curve of the leaf.

12 Using the curve drawn in step 11 as the starting point, draw curved drop shapes.

13 Follow the curve and draw drop shapes accordingly, mind the size and spacing of the drop shapes.

14 Starting from between the petals, draw a downward spiral.

15 Starting from between the petals, draw a downward hook ended waved curve.

16 Starting from the tip of the curve drawn in step 15, draw downward drop shapes.

17 Follow the curve and draw drop shapes accordingly, mind the size and spacing of the drop shapes.

18 Starting from the end of the counterclockwise spiral drawn in step 14, draw a spiral to the right and extending to the ring finger.

19
以藤蔓為起點，往手指畫
出帶勾形的弧線。

20
以步驟 19 的弧線頂端為起
點，往手指畫出弧線，尾
端與弧線尾端密合，成一
水滴狀。

21
在水滴內側畫出直線裝飾。

22
沿著水滴上下的弧度，往
右畫出加長水滴。

23
在水滴頂端，往右擠出圓
點裝飾。

24
沿著步驟 18 的弧度，往右
畫出兩條加長水滴。

25
以花瓣之間為起點，往手
指畫出尾端帶勾形的弧線。

26
以花瓣之間為起點，畫出
弧線。

27
弧線畫至尾端時往下收，
再往外拉出弧度，與波浪
曲線的尾端密合，完成作
品上方的輪廓。

19 Starting from the base of the spiral, draw a hooked curve to the finger.
20 Starting from the tip of the curve drawn in step 19, draw a curve to the finger and meet the end of the curve with the one drawn in step 19 to form a drop shape.
21 Draw a line within the drop shape as shown.
22 Draw an elongated drop shape to the right of the drop shape drawn in step 20 following its' curviness.
23 Draw dots from the tip of the drop shape drawn in step 20 for decoration.
24 Following the curviness of the curve drawn in step 19, draw 2 upward elongated drop shapes.
25 Starting from between the petals, draw a hook ended curve to the finger.
26 Starting from between the petals, draw a curve.
27 Retract the tail of the curve inwards, and then curve outwards and join the end with the end of the curve drawn in step 25 thus finishes the top portion of this design.

28
在輪廓內側畫出弧線。

29
沿著步驟 **28** 的弧度，依
序**畫出圓**形花紋裝飾。

30
沿著圓形花紋的弧度，畫
出弧線裝飾。

31
以顏料填滿輪廓尾端。

32
以花瓣之間為起點，沿著
輪廓上方弧度，畫出加長
水滴。

33
以花瓣之間為起點，畫出
左旋藤蔓。

34
以花瓣之間為起點，畫出
弧線，尾端與左旋藤蔓密
合，完成第二片葉子的輪
廓。

35
以葉子輪廓頂端為起點，
在內側畫出弧線。

36
以直線葉脈依序填滿葉子
左側。

28 Draw a curve in side the contour drawn in step 27.

29 Follow the curviness of the curve drawn in step 28 and drawn on semi-circle patterns as shown.

30 Follow the curviness of the row of semi-circle patterns drawn in step 29, draw on curved lines for decoration.

31 Use the Henna paste to fill in the tail of the contour.

32 Follow the curviness of the upper curve of the contour and draw an elongated drop shape.

33 Starting from between the petals, draw a clockwise spiral.

34 Starting from between the petals, draw a curve and meet the end with the clockwise spiral thus completes the contour of the second leaf.

35 Starting from the base of the second leaf contour, draw a curved line inside the contour.

36 Use straight lines to fill in the left leaf veins.

37
以葉子右側輪廓頂端為起點，在內側往上畫出直線，尾端收尖。

38
在葉子右側畫出螺旋，尾端以水滴收尾。

39
以步驟 37 的螺旋下方為起點，沿著弧度，畫出弧形加長水滴。

40
以葉子尾端為起點，往上畫出加長水滴。

41
以花瓣之間為起點，畫出帶勾形的弧線。

42
以花瓣之間為起點，畫出弧線，尾端與勾形線條密合，完成第三片葉子輪廓。

43
以葉子輪廓頂端為起點，在內側往上畫出直線，尾端收尖。

44
以葉子尾端為起點，在左側往上畫出加長水滴。

45
以葉子尾端為起點，往上畫出長短不一的加長水滴，須帶明顯弧度。

46
最後以第一片葉子下方為起點，依序畫出加長水滴裝飾即可。

37 Starting from the base of the right portion of the second leaf contour, draw a sharp tailed upward line.
38 Draw a spiral in the upper left direction following the leaf's right profile, end in a drop shape matter.
39 Starting from below the spiral drawn in step 37, following the curviness of the spiral, drawn a curved elongated drop shape.
40 Starting from the base of the leaf, draw an elongated drop shape.
41 Starting from between the petals, draw a hooked curve.
42 Starting from between the petals, draw another curve and join the end with the curve drawn in step 41 thus finishes the contour of the third leaf.
43 Starting from the base of the third leaf contour, draw an upward sharp tailed line.
44 Starting from the base of the third leaf contour, draw an elongated drop shape along the leaf's left profile.
45 Starting from the base of the third leaf contour, draw 2 uneven elongated drop shapes along the leaf's right profile.
46 Lastly, starting from the base of the first leaf, draw elongated drop shapes for decoration to complete the pattern.

肩部 Henna —
印度羽飾
Henna on the shoulder —
The Indian Feathered Design

步驟説明

1
先擠出一滴顏料，在構圖下方畫出勾形在下的變形蟲輪廓。

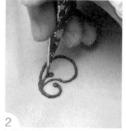

2
以變形蟲的密合處為頂點，於輪廓內側往右上畫出水滴。

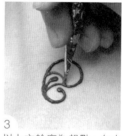

3
以上方輪廓為起點，在水滴右側由上往下畫出弧線。

4
沿著步驟 3 的弧度，在右側畫出平行略長的弧線。

5
依序以直線填滿平行線內的空間，須注意間隔與排列弧度。

6
沿著步驟 4 的弧度，在右側畫出弧線。

7
在步驟 6 的弧線上依序擠出圓點裝飾，須注意大小與間隔。

8
以變形蟲輪廓尾端為起點，往上畫出右旋藤蔓。

Step by step

1 Squeeze a drop of Henna pasted and draw an amoeba contour with a downward hook in the lower portion of the construction.
2 Starting from where the lines connect, draw a drop shape in the upper right direction within the contour.
3 Drawing downward from the top of the contour, draw a vertical curve to the right of the drop shape.
4 Draw on the right another slightly longer vertical curve parallel to the one drawn in step 3.
5 Use striped lines to fill in the space between the parallel lines, take note of the spacing and curviness alignment.
6 Draw on the right another vertical curve parallel to the one drawn in step 4.
7 Draw a row of dots onto the curve drawn in step 6, and mind the size and spacing.
8 Starting from the mid-point on the right side of the amoeba contour, draw an upward counterclockwise spiraled vine.

9
以變形蟲輪廓尾端為起點，往上畫出弧線，以水滴收尾。

10
以變形蟲輪廓尾端為起點，在步驟 9 的弧線左側往上畫出略短弧線，以水滴收尾。

11
以變形蟲輪廓尾端為起點，沿著變形蟲上半部的弧度，在步驟 10 的弧線下方畫出弧線，以朝上的水滴收尾。

12
以步驟 10、11 的弧線之間為起點，沿著步驟 11 的弧度，往下依序畫出橢圓圖案。

13
將橢圓圖案畫至弧線尾端，須注意大小與排列弧度。

14
以變形蟲上方的右旋藤蔓右側為起點，往上畫出右旋藤蔓。

15
以步驟 14 的右旋藤蔓為起點，往右下畫出水滴。

16
以變形蟲輪廓與步驟 14 的水滴之間為起點，往內畫出加長水滴。

17
重複步驟 16，往右依序畫出加長水滴，須注意大小與排列弧度。

9　Starting from the mid-point on the right side of the amoeba contour, draw a curved line with a drop shaped end in the upper direction.

10　Starting from the mid-point on the right side of the amoeba contour, draw a slightly shorter curved line with a drop shaped end left to the curved line drawn in step 9 as shown.

11　Starting from the mid-point on the right side of the amoeba contour, draw a drop shape ended curved line under the one drawn in step 10.

12　Starting from in between the curves drawn in steps 10 and 11, draw downward oval shapes along the curve drawn in step 11.

13　Finish the row of oval shapes to the end of the curve, and take note of the size and curviness alignment.

14　Starting from the mid-point on the right side of the amoeba contour, draw an upward counterclockwise spiraled vine.

15　Starting from the counterclockwise spiraled vine drawn in step 14, draw a downward drop shape.

16　Starting from in between the downward drop shape drawn in step 14 and the amoeba contour, draw an elongated drop shape to the left.

17　Repeat step 16, and draw more elongated drop shapes in between the one drawn in step 15 and the one drawn in step 16; take note of the size and curviness alignment.

18
以變形蟲上方的右旋藤蔓
為起點,往上畫出波浪曲
線,以水滴收尾。

19
沿著步驟 18 的弧度,在右
側往上畫出波浪曲線。

20
將步驟 19 的曲線尾端收成
螺旋。

21
沿著步驟 19 的弧度,以右
旋藤蔓頂端為起點,往上
畫出弧線,以水滴收尾。

22
沿著右旋藤蔓的弧度,依
序畫出弧線,須注意長度
與間隔。

23
以步驟 18、20 的曲線間隔
為起點,往上畫出波浪曲
線,以水滴收尾。

24
沿著步驟 23 的弧度,在左
側往上畫出略短的弧線,
以水滴收尾。

25
最後重複步驟 24,畫出更
短的弧線,即可完成作品。

18 Draw a waved curve with a drop shaped tail extending upward starting from the starting point of the counterclockwise spiraled vines.

19 Follow the curviness of the curve drawn in step 18, and draw a waved curve extending in the upper right direction.

20 Extend the tail of the curve drawn in step 19 into a spiral.

21 Starting from the top of the counterclockwise spiraled vine, draw drop shape ended curves from the bottom-up following the curviness of the curve drawn in step 19.

22 Draw drop shape ended curves along the counterclockwise spiraled vine's curve, and mind the length and spacing of each curve.

23 Starting from in between the curves drawn in steps 18 and 20, draw an upward drop shape ended waved curve.

24 Following the curviness of the curve drawn in step 23, draw a slightly shorter drop shape ended curve in the upper left corner.

25 Repeat step 24 and draw an even shorter curve to complete the design.

頸部 Henna —

小蘭花

Henna on the neck —
The Orchid Design

步驟説明

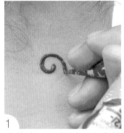

1
先擠出一滴顏料，畫出第一個螺旋。

2
將第一個螺旋尾端往下拉出線條。

3
從線條尾端畫出弧線。

4
將右上弧線往下彎成圓弧。

5
將圓弧線往左下拉，連接尾端，畫出第一個延伸型水滴。

6
從延伸型水滴尾端拉出直線，畫出第一片橢圓葉形。

7
在螺旋與第一片橢圓葉的空隙，往上拉出弧線。

8
將弧線拉至需要長度，頂端往下彎成圓弧。

Step by step

1 Squeeze out a drop of Henna paste and draw a spiral.
2 Extend the tail of the spiral downward into a line.
3 Then, turn upward and extend the tail into a curve in the upper right direction.
4 Make a downward u-turn and create a rounded curve.
5 Extend the curve in the lower left direction, then connect it with the bottom end, which creates the first elongated drop contour.
6 Draw a line within the contour starting from the base, thus finishes the first oval shaped leaf.
7 Starting from in between the first oval shaped leaf and the spiral, extend a curve.
8 Extend the curve as needed, and turn downward to create a rounded curve.

9 將步驟 8 的圓弧畫成第二個螺旋，再將筆尖提起收尾。

10 如果繪圖時產生斷線，可擠出少許顏料修補。

11 沿著第一個螺旋頂端，往右拉出弧線。

12 將步驟 11 的弧線拉至第一、二個螺旋中間，再往右上拉，沿著第二個螺旋頂端拉出弧線。

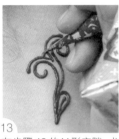
13 在步驟 12 的 V 形空隙，以弧線畫出花萼。

14 在花萼上方畫出圓形花蕊。

15 依序畫出三條花蕊，須注意大小與排列位置。

16 以第一、二條花蕊之間為起點，往左畫出第一片心形花瓣。

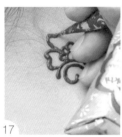
17 以第一、二條花蕊之間為起點，往右畫出第二片心形花瓣。

18 以第三條花蕊頂端為起點，往右畫出第三片心形花瓣。

19 在心形花瓣內側，點出圓點裝飾。

20 依序在心形花瓣上點出三個圓點，再於第一片橢圓葉上點出圓點。

9　Turn the rounded curve in step 8 into the second spiral, and lift the applicator to finish the line.

10　If broken lines were to appear, a small amount of paste may be further applied on the blank area to fix the line.

11　Following the top of the first spiral, draw a curved line to the right.

12　Extend the curve drawn in step 11 downward to the mid-point between the 2 spirals, and then extend the line upward to create another curve above the second spiral.

13　Within the just drawn V-shape from step 12, draw 2 curved lines to form the calyx.

14　Draw rounded pistils on top of the calyx.

15　Draw a row of 3 pistils, and mind the size and alignment of the pistils.

16　Starting from between the first and second pistil, draw the first heart shaped petal to the left.

17　Starting from between the first and second pistil, draw the second heart shaped petal to the right.

18　Starting from the top of the third pistil, draw the third heart shaped petal to the right.

19　Draw dots within the petal for decoration.

20　Finish dotting the 3 petals, and then draw another dot above the oval shaped leaf as shown.

21
將第一片橢圓葉上的圓點往下拉出第一滴加長水滴，尾端須連接於橢圓葉內側。

22
重複步驟 **20-21**，在第一滴加長水滴右側畫出第二滴加長水滴。

23
在第一片橢圓葉左側，往左畫出第二個延伸型水滴。

24
從延伸型水滴的尾端拉出直線，畫出第二片橢圓葉形。

25
在第二片橢圓葉形上方點出圓點。

26
將圓點往下拉出第三滴加長水滴，尾端須連接於橢圓葉內側。

27
重複步驟 **25-26**，在第三滴加長水滴右側依序畫出第四、五滴加長水滴。

28
在第一、二片橢圓葉尾端，往右拉出弧線。

21 Extend the dot above the oval shaped leaf downward into the first elongated drop shape, the tail of the drop shape should end in the corner between the stem and the first oval shaped leaf.

22 Repeat steps 20 and 21, and draw a second elongated drop shape to the right of the first one.

23 To the left of the first oval shaped leaf, draw a second elongated drop contour to the left.

24 From the base of the elongated drop shape, extend a line and form the second oval shaped leaf.

25 Draw a dot above the second oval shaped leaf.

26 Extend the dot downward into the third elongated drop shape, the tail of the drop shape should end in the corner between the stem and the second oval shaped leaf.

27 Repeat steps 25 and 26, and draw a fourth and fifth elongated drop shape to the right of the third elongated drop shape.

28 From the base of the oval shaped leaves, extend a curved line to the right.

29
將步驟 28 的弧線往內畫
出第三個螺旋。

30
在第一片橢圓葉與第三個
螺旋的 V 形空隙，往右拉
出弧線。

31
將步驟30的弧線往上拉，
須注意維持弧度。

32
在步驟 30 的弧線尾端擠出
圓點，畫出第六滴加長水
滴。

33
重複步驟 30-32，在第六
滴加長水滴右側畫出稍長
的第七滴加長水滴。

34
在第二片橢圓葉下方，往
左畫出第四個螺旋。

35
在第二片橢圓葉與第四個
螺旋的 V 形空隙，往左拉
出弧線。

36
最後在步驟 35 的弧線尾
端擠出圓點，畫出第八滴
加長水滴，即可完成作品。

29　Retract the tail of the curve drawn in step 28 to form the third spiral.
30　Draw a curved line to the right starting from the V-shaped corner between the first oval shaped leaf and the third spiral.
31　Extend the tail of the curve drawn in step 30 upward, take note of the curviness.
32　Then, add a rounded dot at the end of the curve drawn in step 30, and create the sixth elongated drop shape.
33　Repeat steps 30 through 32, and draw a seventh and eighth elongated drop shape to the right of the third elongated drop shape.
34　Starting from under the second oval shaped leaf, draw the fourth spiral to the left.
35　Draw a curved line to the left starting from the V-shaped corner between the second oval shaped leaf and the fourth spiral.
36　Lastly, add a rounded dot at the end of the curve drawn in step 35, and create the eighth elongated drop shape to complete the design.

胸部 Henna ——
小鳳凰
Henna on the chest ——
Phoenix Totoo Design

步驟説明

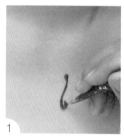

1
先擠出一滴顏料，在構圖
的最左側，往左下畫出帶
勾形弧線。

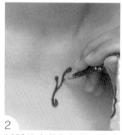

2
以弧線中段為起點，往右
上畫出弧線，以水滴收尾，
完成頭羽。

3
以弧線中段為起點，往下
畫出圓弧線。

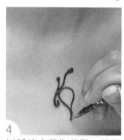

4
以弧線中段為起點，往下
畫出弧線，尾端與圓弧線
密合，完成頭部輪廓。

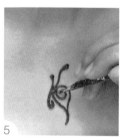

5
以頭羽下方為起點，往下
畫出螺旋。

6
沿著螺旋的弧度，在下方
畫出弧線，完成鳥眼。

7
以步驟 6 的弧線為起點，
往下畫出直線，完成鳥嘴
輪廓。

8
以頭羽右側為起點，往右
畫出藤蔓。

Step by step

1 Squeeze a drop of Henna paste and draw a downward hook ended curve to the far left of the construction.
2 Starting from the mid-point of the previous curve line, draw a drop shape ended curve in the upper right direction to finish the head feather.
3 Starting from the mid-point of the first curve, draw a downward rounded curve.
4 Starting from the mid-point of the first curve, draw another downward curve and meet the end with the one drawn in step 3 to finish the contour of the phoenix's head.
5 Draw a spiral at the base of the head contour as shown.
6 Following the curviness of the spiral, draw a curve under it to finish the eye of the phoenix.
7 Starting from the curve drawn in step 6, draw a downward line to complete the contour of the beak.
8 Starting from the right side of the head feather, draw a spiraled vine to the right.

9
沿著藤蔓的弧度，往下畫
出圓弧線，以水滴收尾。

10
沿著步驟 9 的弧度，往下
畫出略長的圓弧線，以水
滴收尾。

11
在步驟 9 的弧線上方，擠
出小圓點裝飾。

12
沿著步驟 8 的藤蔓弧度，
依序擠出小圓點，須注意
大小與間隔。

13
以步驟 10 的圓弧線與鳥眼
的間隔為起點，往下畫出
加長水滴。

14
在步驟 13 的水滴右側，往
右下畫出略長的加長水滴。

15
在步驟 14 的水滴右側，往
右畫出加長水滴，完成鳥
的身體。以上須注意間隔
與排列弧度。

16
以步驟 10 的弧線尾端為起
點，往下畫出小圓弧線。

17
以小圓弧線中段為起點，
往上畫出波浪曲線，尾端
收成螺旋。

9 Following the curviness of the spiraled vine, draw a rounded curve with a drop shaped end under it.
10 Following the curviness of the curve drawn in step 9, draw a slightly longer rounded curve with a drop shaped end under it.
11 Draw a row of dots for decoration starting from the tip of the curve drawn in step 9.
12 Draw dotted patterns along and above the spiraled vine from the right to the left; take note of the size and spacing of each dot.
13 Starting from the point in between the rounded curve drawn in step 10 and the phoenix's eye, draw a downward elongated drop shape.
14 To the right of the elongated drop shape drawn in step 13, draw another slightly longer elongated drop shape in the lower right direction.
15 Starting from the right of the elongated drop shape drawn in step 14, extend an elongated drop shape to the right to compete the phoenix's body. Remember to take note of the spacing and curviness alignment of each elongated drop.
16 Starting from the end of the curve drawn in step 10, draw a small downward rounded curve.
17 Starting from the mid-section of the small rounded curve, draw a waved curve in the upper direction, and extend the tail into a spiral as shown in the diagram.

18
沿著波浪曲線的弧度，往
右上畫出略長的波浪曲線，
以水滴收尾。

19
以步驟 **17**、**18** 的間隔為起
點，往上畫出加長水滴。

20
以步驟 **18** 的曲線頂端為起
點，往上畫出水滴裝飾。

21
沿著曲線弧度，依序畫出
水滴，須注意大小與排列
方式。

22
以步驟 **17** 的曲線底部為起
點，畫出水滴裝飾。

23
沿著曲線弧度，依序畫出
水滴，須注意大小與排列
方式。

24
以步驟 **17** 的螺旋頂端為起
點，往右上畫出圓弧線。

25
最後沿著圓弧線弧度，依
序畫出漸漸縮短的弧線完
成尾羽，即可完成作品。

18 Following the curviness of the waved curve, draw another drop shape ended waved curve in the upper right direction.
19 Starting from the point in between the curves drawn in steps 17 and 18, draw an upward elongated drop shape.
20 Starting from the starting point of the curve drawn in step 18, draw drop shapes.
21 Follow the curviness of the curve and draw on a row of drop shapes, take note of the size and alignment of each shape.
22 Starting from the ending point of the curve drawn in step 17, draw drop shapes for decoration.
23 Follow the curviness of the curve and draw on a row of drop shapes, take note of the size and alignment of each shape.
24 Starting from the top of the spiral drawn in step 17, draw an upward rounded curve.
25 Following the curviness of the rounded curve, draw 2 gradually shorted curves under it, thus finishes the tail feathers and completes the design.

Body Henna 小祕訣

上油
Mehndi oil application

步驟説明：進行 Henna 的人體彩繪前，可以 Mehndi 油塗抹要作畫的部位，讓顯色更持久。

Instructions：Apply the Mehndi oil onto the area prior to Henna paste application, for a long-lasting Henna design.

 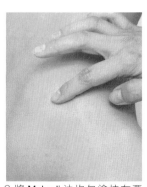

1 取適量 Mehndi 油倒在手上。
Pour an appropriate amount of Mehndi oil on your hand.

2 將 Mehndi 油均勻塗抹在要作畫的部位。
Apply the Mehndi oil evenly on the area which Henna will be drawn on.

噴水
Water spraying

步驟説明：在顏料未剝落前，可在圖案上噴水，延緩剝落時間，讓圖騰的輪廓與顯色更明顯。

Instructions：Spray water on the design before the Henna paste dries, to postpone the peeling of the paste and create a design with more defined lines and brighter color.

1 取噴霧式水瓶。
Take the spray bottle.

2 將水均勻噴灑在輪廓上即可。
Evenly spray the water onto the Henna design.

Body Henna 小祕訣

剝落
Peeling

  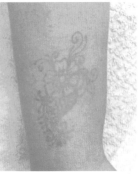

1 先觀察顏料是否乾透，乾透的顏料會有細小裂紋。
First observe whether or not the paste has completely dried, there should be fine cracks present.

2 顏料乾透後會自行脫落，或用手指輕輕剝落亦可。
The dried paste will peel off on its' own, or you could also use your fingers to assist the peeling process.

3 顏料剛脫落的圖騰顯色較淺，1-3天後，會依個人狀況顯現出不一樣的顏色。
Freshly peeled designs may seem lighter in color, and will show different coloring results depending on the individual condition after 1 to 3 days of application.

小叮嚀：
　　Henna 顏料在皮膚上的乾燥時間長短、顯色狀況會依當時所處環境的溫濕度、個人體質而略顯不同。例如，雨天或所處環境濕度較高時，顏料所需的乾燥時間較長；反之，待在冷氣房或所處環境較乾燥時，顏料所需的乾燥時間則縮短。
　　若以個人體質來說，健康狀況較好、體溫較高的人，顏料乾燥速度越快，顯現的顏色也會越美麗；如果本身體質較虛弱，或體溫較低，顏料乾燥的速度會比較慢，顯色的狀況與持久度也較一般體質差。

Gentle reminder:
　　To prolong Henna paste stays on the skin will depends on the drying duration, temperature, humidity and also Individual body temperature. For example, the Henna paste needs more time to dry up during raining day; but if stays in an air-conditioned room or in dry environment the drying duration of henna paste will be shorten.
　　The Henna paste will dry up faster, colour will stay longer and more obvious according to personal physical ability, healthy and higher body heat temperature; If individual physical is weak, or body temperature is low, the duration of drying will be longer and slower, also the colour appearance will not that obvious and not so sustainable.

作品欣賞

作品欣賞

作品欣賞

Henna Art First experience —
understand the art of hand painted in India

書　　　名	Henna Art 初體驗：認識印度的手繪藝術
作　　　者	小美
主　　　編	昕彤國際資訊企業社・陳姵君
美　　　編	昕彤國際資訊企業社・呂宛儒
封面設計	洪瑞伯
攝影師	吳曜宇、Eva
模特兒	Maggie
發行人	程顯灝
總編輯	盧美娜
美術編輯	博威廣告
製作設計	國義傳播
發行部	侯莉莉
財務部	許麗娟
印　　　務	許丁財
法律顧問	樸泰國際法律事務所許家華律師
藝文空間	三友藝文複合空間
地　　　址	106 台北市安和路 2 段 213 號 9 樓
電　　　話	（02）2377-1163
出版者	四塊玉文創有限公司
總代理	三友圖書有限公司
地　　　址	106 台北市安和路 2 段 213 號 9 樓

電　　　話	（02）2377-4155、（02）2377-1163
傳　　　真	（02）2377-4355、（02）2377-1213
E - m a i l	service @sanyau.com.tw
郵政劃撥	05844889 三友圖書有限公司
總經銷	大和書報圖書股份有限公司
地　　　址	新北市新莊區五工五路 2 號
電　　　話	（02）8990-2588
傳　　　真	（02）2299-7900

初版　2023 年 9 月
定價　新臺幣 368 元
ISBN　978-626-7096-54-3（平裝）

國家圖書館出版品預行編目（CIP）資料

Henna Art初體驗：認識印度的手繪藝術 / 小美作.
-- 初版. -- 臺北市：四塊玉文創有限公司, 2023.09
　　面；　公分
　中英對照
　ISBN 978-626-7096-54-3（平裝）

1.CST: 人體彩繪 2.CST:印度

425　　　　　　　　　　　　　　　112014267

三友官網　　　三友 Line@